Smart Agriculture

Volume 1

The book series Smart Agriculture presents progress of smart agricultural technologies, which includes, but not limited to, specialty crop harvest robotics, UAV technologies for row crops, innovative IoT applications in plant factories, and big data for optimizing production process. It includes both theoretical study and practical applications, with emphasis on systematic studies. AI technologies in agricultural productions will be emphasized, consisting of innovative algorithms and new application domains. Additionally, new crops are emerging, such as hemp in U.S., and covered as well. This book series would cover regions worldwide, such as U.S., Canada, China, Japan, Korea, and Brazil.

The book series Smart Agriculture aims to provide an academic platform for interdisciplinary researchers to provide their state-of-the-art technologies related to smart agriculture. Researchers of different academic backgrounds are encouraged to contribute to the book, such as agriculture engineers, breeders, horticulturist, agronomist, and plant pathologists. The series would target a very broad audience – all having a professional related to agriculture production. It also could be used as textbooks for graduate students.

More information about this series at http://www.springer.com/series/16785

Zhao Zhang · Zhaohua Zhang ·
Cannayen Igathinathane · Yingkuan Wang ·
Yiannis Ampatzidis · Gang Liu
Editors

Mechanical Harvest of Fresh Market Apples

Progress over the Past Decades

Springer

Editors
Zhao Zhang
Key Lab of Modern Precision Agriculture
System Integration Research
Ministry of Education
China Agricultural University
Beijing, China

Cannayen Igathinathane
Department of Agricultural and Biosystems
Engineering
North Dakota State University
Fargo, ND, USA

Yiannis Ampatzidis
University of Florida
Immokalee, FL, USA

Zhaohua Zhang
College of Economics and Management
Shandong Agricultural University
Tai'an, Shandong, China

Yingkuan Wang
Chinese Academy of Agricultural
Engineering
Beijing, China

Gang Liu
MOE Key Laboratory of Modern Precision
Agriculture System Integration Research
China Agricultural University
Beijing, China

ISSN 2731-3476 ISSN 2731-3484 (electronic)
Smart Agriculture
ISBN 978-981-16-5315-5 ISBN 978-981-16-5316-2 (eBook)
https://doi.org/10.1007/978-981-16-5316-2

This Springer imprint is published by the registered company Springer Nature Singapore Pte Ltd.
The registered company address is: 152 Beach Road, #21-01/04 Gateway East, Singapore 189721, Singapore

Series Book Editor Preface

Rapid technology progress in terms of new sensors, Internet of Things (IoT), robotics, drones, big data, and innovative artificial intelligence (AI) algorithms, is profoundly transforming agriculture. Driven by these new technologies, the traditional agriculture manner featured by low efficiency, high labor intensity, environmental pollution, and unsustainability is gradually updated to high-efficient, unmanned, minimal pollution, and sustainable. The evolution of smart agriculture is a key to feed the increasing population in a sustainable and environmentally friendly manner. Though thousands of projects in the domain of smart agriculture are conducted with outcomes published, there lacks a book series to systematically and holistically to present the technology progress, which drives the organization of the book series titled "Smart Agriculture" by collaborating with Springer Nature.

As the first volume of this book series, this book specially focuses on the technology progress in mechanical harvest of fresh market apples. The ladder–bucket apple harvest method has been dominant for decades, and due to the pressure of high labor cost, harvest labor shortage, and increasing market demand, researchers and practical engineers have invested tremendous efforts to realize mechanical harvest of fresh market apples, including bulk harvest, harvest robotics, and harvest-assist platform, which consist of the contents of this book. The book presents the technology progress of apple harvest during the past 70 years and points the existing bottlenecks and future trends. Researchers and undergraduates/graduates can read this book to quickly obtain the hints on the specific topic of the history of mechanical harvest of fresh market apples.

Starting from this book, I am about to organize several others books, such as "UAV application in smart agriculture" and "innovative sensing technologies for agriculture products," which would be online soon. I am trying my best to catch up the rapid development pace of smart agriculture and have the research outcomes published in this book series.

Beijing, China Zhao Zhang

Preface

The current ladder–bucket method has been the dominant fresh market apple harvest approach during the past decades. Workers pick low-level apples by standing on the ground while taking advantage of ladders to get access to high-level fruits. The ladder–bucket method has a myriad of disadvantages, which include, but are not limited to, inefficiency, high-cost, demanding strength requirements, and occupational hazards. With increasing labor cost, shrinking agricultural labor pool, and tight immigration policies, US apple industry faces challenges to be competitive both nationally and internationally. Mechanical harvest of apples is a potential technology candidate to replace the ladder–bucket method and address all these challenges. In fact, research on developing mechanical harvest of fresh market apples started from the middle of twentieth century. A huge number of projects were completed during 1950s to 1980s, and then suddenly all research stopped due to the government halting financial support, as the government worried that the newly developed technology may result in high unemployment rates for agriculture employees. Then, at the end of twentieth century and beginning of twenty-first century, technology boom in terms of new machine/deep learning algorithms, big data and high-speed computation hardware, coupled with the fact that apple growers were unable to find sufficient seasonal migrant employees to harvest apples, the government restarted to invest large grants to develop innovative mechanical fresh market apple harvest technologies.

Beyond harvest, apple postharvest handling (e.g., storage, grading, sorting, and packing) is the other major contributing factor for the high production cost of the US apple industry. Though it is known, validated, and accepted by apple growers that infield sorting would be beneficial, few studies have been conducted. Very recently, researchers of the U.S. Department of Agriculture, Agricultural Research Service developed, improved, validated, and demonstrated an apple infield sorting system. Furthermore, the group integrated the infield sorting system into a customized apple harvest platform, which resulted in the first apple harvest and infield sorting machine. Multi-year field tests demonstrated the satisfactory performance of the machine.

Though a huge number of projects completed and hundreds of papers, patents, and magazine articles published during the past decades, there lacks a book that holistically and systematically provides an introduction to the technology evolution, which drives the idea of drafting this book. The first four chapters of this book focus on the apple harvest-assist unit, and the second four chapters on the apple harvest and infield sorting machine. Chapter 1 reviews the history of technology progress on mechanical harvest of fresh market apples, including bulk harvest, robots, and harvest platforms. The bulk harvest method was so extensively tested starting in the 1950s and continuing on the following decades, and it finally does not end in commercial application due to unacceptable apple bruising. Research on harvest robots started from early 1990s, and recently, several startups are on their way to commercialize their apple harvest robots. However, there are a huge number of technological bottlenecks ahead before the developed robots being in commercial stage. Due to the unsuccessful results of bulk harvesting and robots, researchers and practical engineers shifted to the direction in developing harvest platforms. The adoption of platforms for apple harvest is superior to the ladder–bucket method in terms of high harvest efficiency and low incidence on occupational injuries. More importantly, the platform would avoid the occurrence of ladder fall accidents (would easily lead to fracture and death) associated with the conventional ladder–bucket method. Following the Chaps. 1 and 2 describes the development and improvement of a low-cost apple harvest-assist unit, with its economic and ergonomic evaluations provided in Chaps. 3 and 4, respectively. Chapter 5 presents the development and test of two apple sorters, which finally satisfactorily met the requirement of processing nine apples. Before integrating the apple harvest and infield sorting machine, Chap. 6 describes an innovatively developed bin filler, a mechanism to catch apples and then deliver them into the bin gently with minimal bruising, that meets the requirement to work seamlessly with the apple sorter. After integrating the apple harvest and infield sorting machine, Chap. 7 contains an economic analysis to show apple growers the benefits of adopting the technology. Moreover, dealers need the economic analysis results when they introduce the harvest and infield sorting machine to apple growers. To find further efficiency improvement room for the practically applied apple harvest methods (i.e., ladder–bucket and different harvest platforms), a time and motion study for evaluation of apple harvest processes with different harvest methods was conducted, which comprises Chap. 8.

I hope this book can provide a quick hint to relevant researchers and those who are interested in apple harvest technology some valuable information to understand the technology trend quickly and comprehensively.

Beijing, China Zhao Zhang
Tai'an, China Zhaohua Zhang

Contents

Editors and Contributors

About the Editors

Dr. Zhao Zhang is currently working with Key Lab of Modern Precision Agriculture System Integration Research, Ministry of Education, China Agricultural University, China. His major research direction relates to sensing and automation in agriculture, focusing on applying and developing innovative technologies (e.g., UAVs and ground vehicle-based sensors) to support sustainable agriculture. Projects going on include, but are not limited to, using drone imagery for automatic crop disease detection and growth condition monitoring. In addition, he collaborates with a startup on developing an automatic rock picker. One project is to develop a proximal sensing system to recognize rocks and then to guide an end-effector to the target rock. Before joining NDSU, he worked in the USDA Agricultural Research Service (ARS) Sugarbeet and Bean Research Unit at East Lansing, Michigan. His research focused on the development of innovative engineering technologies for harvesting and automated grading and sorting of apples in the orchard. He was primarily involved with system integration, as well as automatic control design and implementation. His research interests also included cost-benefit analysis of adopting mechanical harvest aid/sorting machines. He was a co-inventor for the infield sorting system (US Patent 9,919,345). In addition, he worked on the development and integration of an innovative apple harvest robot. Before joining USDA/ARS, he had completed his Ph.D. studies in the

Department of Agricultural and Biological Engineering at The Pennsylvania State University. His Ph.D. research was focused on developing a low-cost apple harvest-assist unit to improve labor productivity and decrease pickers' occupational injuries.

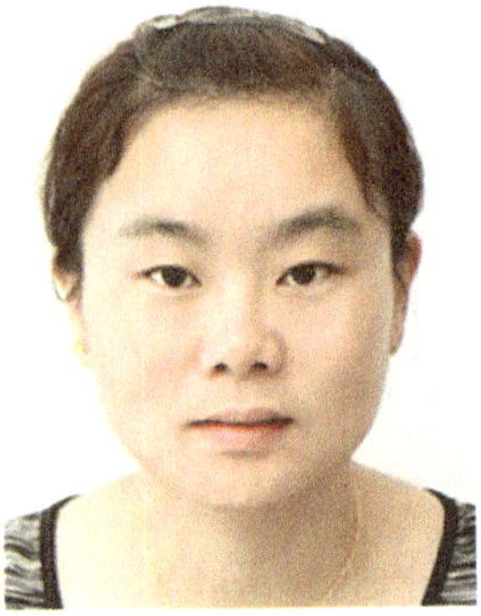

Dr. Zhaohua Zhang is an associate professor at the College of Economics and Management, Shandong Agricultural University, China. Dr. Zhang earned her Ph.D. from Department of Agricultural Economics and Rural Sociology, Auburn University, USA. Her Ph.D. research evaluated impacts of environmental equity (e.g., equity of air pollution, heavy metal contamination of land, and toxic chemicals released in water) on households of different races using multiple econometric models based on household residential choices in the face of these disamenities. After graduating from Auburn University, she joined Shandong Agricultural University, China, as an assistant professor. Currently, her research areas relate to agricultural economics and rural development. Special focus is given to willingness of farmers to adopt innovative technologies in agricultural production, and how these technologies could improve local environmental quality and increase farmers' income, applying econometric modeling strategy with big data. In addition, she also works on welfare analysis of different policy interventions in agriculture sector using counterfactual simulations, especially evaluates how policies for agricultural material subsidies, farm machinery purchase subsidy, and seed subsidies affect economic status of the rural population. Dr. Zhang also works on economic analysis of agricultural machinery, which would provide farmers baseline information for decision making on the investment.

Dr. Cannayen Igathinathane is an associate professor in the Department of Agricultural and Biosystems Engineering, North Dakota State University (NDSU), Fargo, ND, USA. He currently focuses on engineering-related issues of biomass feedstock preprocess engineering including logistics and precision agriculture using image processing and tool development. Specific research interests include biomass physical, mechanical, and thermal quality analysis; postharvest and agricultural process and food engineering; mathematical modeling and numerical simulation; energy beet front end processing; utilization of damaged wood biomass; image processing, machine vision, user-coded software tool development, computer applications in agricultural engineering using open-source software; and innovation in research. He collaborates research and other outreach activities with scientists of Northern Great Plains Research Laboratory, USDA-ARS, Mandan, ND, and others in the state and the nation. He was earlier involved in biomass feedstock engineering and related fields at the University of Tennessee, TN; Mississippi State University (MSU), MS; University of British Columbia, Canada; and Andhra Pradesh Agricultural University, India. Dr. Cannayen has authored several technical peer-reviewed original articles (75), book chapters (5), laboratory manuals (2), textbook (1), and posters/presentations (>210). He is also well recognized among the scientific community nationally and internationally and has served as a reviewer for more than 400 articles from scientific journals and technical proposals from several national and international organizations. He has developed several innovative methodologies in the field of agricultural and biomass process engineering and open-source image processing applications in agriculture (https://www.ndsu.edu/aben/faculty_and_staff/personnel/cannayen/). Dr. Cannayen is a member of ASABE, ISAE, Gamma Sigma Delta, and editorial board member of "Journal of Renewable Energy" and "Scientifica" peer-reviewed journals. He was also the recipient of significant awards, such as several Outstanding Reviewer Awards (Elsevier journals and ASABE), ASABE Superior Paper (2018), ASABE ITCSC Technical Committee Paper Award (2016), and MSU 2008 Research Support Award. He is currently serving as Principal Investigator of two funded research projects and earned research grants (total $2.45 million) to support research and graduate students (masters and Ph.D.) training.

Dr. Yingkuan Wang is a professor (full-time research) and former Director of Agricultural Engineering Information Center of Chinese Academy of Agricultural Engineering. Dr. Wang is also the Secretary General of Chinese Society of Agricultural Engineering. He was B.S. and M.S. majored Agricultural Mechanization at Northwest Agriculture and Forestry University in 1995 and 1998, respectively. After that, he obtained his Ph.D. of Editing and Publishing at Peking University in 2006, and was a senior visiting scholar at University of Minnesota-St. Paul Campus during 2009–2011. His research interests mainly include agricultural mechanization and automation, agricultural information, digital agriculture, precision agriculture, agricultural engineering, and biomass processing. Prof. Wang has launched an open-access journal, International Journal of Agricultural and Biological Engineering, which has been covered by over 30 international indexing systems including SCI with JIF of 2.032 in JCR2020. And he has been serving as the editor-in-chief of IJABE and Transaction of the CSAE. Dr. Wang has been appointed as adjunct professors of China Agricultural University, Jiangsu University, Shandong University of Technology and Fujian Agriculture & Forestry University, Jiamusi University, etc., serving as editorial board members of eight prestigious journals including four English journals, scholarly members of six research centers and key laboratories. He serves as the Board Member of four international societies and two China national societies in agricultural engineering and editing: CIGR, ASABE, AAAE and AOCABFE, CSAE and CESSP (China Editology Society for Science Periodicals), visiting over 30 countries. Prof. Wang presided over 40 research projects as PI or a major participant and has published over 130 journal articles and ten books or book chapters. Prof. Wang has been appointed as a postdoctoral supervisor with cultivation of five high-quality postdoctors. Prof. Wang received a lot of international and national awards such as AOC Distinguished Service Award, ASABE Presidential Citation Award, ASABE Outstanding International Exchange Award,

China National Silver Ox Award for Outstand Science Editors, Outstanding Contribution Award of China's Agricultural Engineering during the 40-year reform and opening-up (one of the 40 awardees).

Dr. Yiannis Ampatzidis is an agricultural engineer and an Associate Professor in the Agricultural and Biological Engineering Department, University of Florida. He is located at Southwest Florida Research and Education Center (SWFREC). He received his Ph.D. in agricultural engineering from the Aristotle University of Thessaloniki (AUTH, Greece) in 2010, his MS in agricultural engineering in 2005, his B.S. in agricultural engineering (5-year diploma, MSc equivalent) in 2008, and his B.S. in agriculture and ecology (5-year diploma) in 2002 from the same university. He has held several research positions in Greece and USA. Most recently, Dr. Ampatzidis was an associate professor in the Department of Physics and Engineering at the California State University, Bakersfield (CSUB). Before that, he was a researcher in the Center for Precision and Automated Agricultural Systems (CPAAS), at Washington State University (WSU). He has taught at Technological Educational Institute of Thessaloniki, at the Technological Educational Institute of Larisa, and at the AUTH (Greece). Generally, he works in the area of mechanization and automation of specialty crop production, focusing on the design, development, and testing of sensors and control systems for optimal management of inputs, resources, and products. His current research focus is on mechatronics, precision agriculture, smart machinery, machine vision, artificial intelligence, UAVs, and machine systems with special interest in development, implementation, and evaluation of agricultural machines and control systems for high-value crops. He has authored or co-authored more than 100 papers in journals/international conferences, in his areas of expertise.

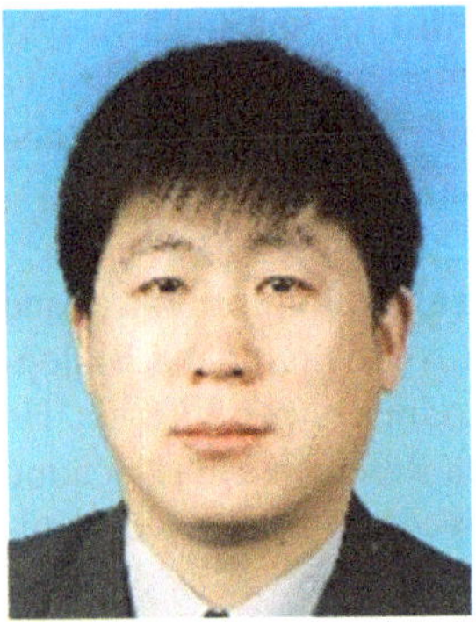

Dr. Gang Liu earned his Ph.D. from China Agricultural University, and his Ph.D. project was methods and experimental research on information processing in precision agriculture. Currently, Dr. Liu is a professor with the College of Information and Electrical Engineering, China Agricultural University, Beijing, China. His major research directions relate to the domain of innovative precision agriculture technology, which include, but are not limited to, Global Navigation Satellite System (GNSS) land intelligent leveling, smart variable fertilizer technology and equipment development, and apple picking robotics. Under the supervision of Dr. Liu, his team developed laser control leveling equipment, and the research output won the 2018 first prize of agricultural water-saving science and technology. So far, Dr. Liu has co-authored more than 90 journal papers, with a majority published in the mainstream journals of agricultural engineering, such as Computer and Electronics in Agriculture, Applied Engineering in Agriculture, and ELECTROCHIMICA ACTA. Beyond that, Dr. Liu has held 38 patents and edited/co-edited seven books.

Contributors

Y. Ampatzidis Agricultural and Biological Engineering Department, Southwest Florida Research and Education Center, University of Florida, IFAS, Immokalee, FL, USA

K. Betitame Department of Agricultural and Biosystems Engineering, North Dakota State University, Fargo, ND, USA

L. Fu College of Mechanical and Electronic Engineering, Northwest A&F University, Yangling, Shaanxi, China

D. Hu College of Optical, Mechanical and Electrical Engineering, Zhejiang A&F University, Hangzhou, Zhejiang, China

M. Li College of Economics and Management, Shandong Agricultural University, Tai'an, China

G. Liu Key Lab of Modern Precision Agriculture System Integration Research, Ministry of Education of China, China Agricultural University, Beijing, China;
Key Lab of Agriculture Information Acquisition Technology, Ministry of Agriculture and Rural Affairs of China, China Agricultural University, Beijing, China

H. Liu Northwest Institute of Eco-Environment and Resources, Chinese Academy of Sciences, Lanzhou, China

W. Lu College of Artificial Intelligence, Nanjing Agricultural University, Nanjing, China

Y. Lu Department of Agricultural and Biological Engineering, Mississippi State University, Starkville, MS, USA

Y. Qiao Australian Centre for Field Robotics (ACFR), Faculty of Engineering, The University of Sydney, Sydney, NSW, Australia

Y. Shi College of Engineering, Nanjing Agricultural University, Nanjing, China

Y. Wang Chinese Academy of Agricultural Engineering, Chinese Society of Agricultural Engineering, Beijing, China;
Agricultural Information Institute of Chinese Academy of Agricultural Sciences, Beijing, China

C. Yang Department of Bioproducts and Biosystems Engineering, University of Minnesota, Saint Paul, MN, USA

Zhao Zhang Key Lab of Modern Precision Agriculture System Integration Research, Ministry of Education of China, China Agricultural University, Beijing, China;
Key Lab of Agriculture Information Acquisition Technology, Ministry of Agriculture and Rural Affairs of China, China Agricultural University, Beijing, China;
Department of Agricultural and Biosystems Engineering, North Dakota State University, Fargo, ND, USA

Zhaohua Zhang College of Economics and Management, Shandong Agricultural University, Tai'an, China

Zhou Zhang Department of Biological Systems Engineering, University of Wisconsin-Madison, Madison, WI, USA

Chapter 1
Technology Evolvement in Mechanical Harvest of Fresh Market Apples

D. Hu, Y. Ampatzidis, G. Liu, Zhao Zhang, and K. Betitame

Abstract Traditional ladder-bucket apple harvest method is inefficiency, prone to causing occupational injuries, and high strength demanding. During the past decades, researchers explored different approaches to realize mechanical harvest of fresh market apples to replace the ladder-bucket approach. This study summarizes the technology progress in terms of bulk harvesting, harvest robots, and harvest platforms. The bulk harvesting concept takes advantage of shaking, pressing, or air jet to automatically detach multiple apples from tress simultaneously. However, due to apple severe bruising conditions, the bulk harvesting method cannot be commercially used to harvest fresh market apples. After detecting and localizing apples from trees, the end effector of a harvest robot would approach the target apple to complete the picking process. Though two apple harvest robots are on their way for commercialization, both have a number of bottlenecks to fix before their commercial applications, such as how to pick apples grown in cluster without bruising. Harvest

D. Hu
College of Optical, Mechanical and Electrical Engineering, Zhejiang A&F University, Hangzhou 311300, Zhejiang, China
e-mail: hudong538338@zju.edu.cn

Y. Ampatzidis
Agricultural and Biological Engineering Department, Southwest Florida Research and Education Center, University of Florida, IFAS, 2685 SR 29 North, Immokalee, FL 34142, USA
e-mail: i.ampatzidis@ufl.edu

G. Liu · Z. Zhang (✉)
Key Lab of Modern Precision Agriculture System Integration Research, Ministry of Education of China, China Agricultural University, Beijing 100083, China
e-mail: zhang.zhao.1@ndsu.edu

G. Liu
e-mail: pac@cau.edu.cn

Key Lab of Agriculture Information Acquisition Technology, Ministry of Agriculture and Rural Affairs of China, China Agricultural University, Beijing 100083, China

Z. Zhang · K. Betitame
Department of Agricultural and Biosystems Engineering, North Dakota State University, Fargo, ND 58102, USA
e-mail: kelvin.betitame@ndsu.edu

Z. Zhang et al. (eds.), *Mechanical Harvest of Fresh Market Apples*, Smart Agriculture 1,
https://doi.org/10.1007/978-981-16-5316-2_1

platforms provide an alternative solution to the ladder-bucket method, and several platforms are already on the market. However, the adoption rate is low because the machine price is high ($50,000~$120,000) and growers are uncertain about the economic benefits generated by technology adoption. Thus, future research should focus on addressing the bottlenecks of the harvest robots, and developing low-cost harvest platform. In addition, more functions should be added to the platform, such as infield sorting, to allow it to bring more benefits by technology adoption, which could help improve the adoption rate.

Keywords Fresh market apples · Bulk harvest · Harvest robot · Harvest platform · Mechanization

1.1 Introduction

As one the most popular fruits worldwide, apples are consumed in a number of manners, such as fresh eating, juice, and sauce [1]. In addition to providing nutrients (e.g., protein and sugar) and non-nutrients (e.g., dietary fiber and vitamins), apples are rich of polyphenols that are beneficial to human health [2]. Fresh market apples are still manually harvested, and due to the inefficiency and high labor cost, harvest cost accounts for more than 25% of the apple total production cost [3]. During the harvest season, workers conduct harvest activity using the ladder-bucket method—for the low level fruit, workers stand on the ground to pick apples, for the high level apples, workers stand on the ladder to complete the harvest. During the entire harvest process, workers need to wear a bucket to temporarily hold apples, and when the bucket is full, workers need to walk to a bin to unload fruit [4]. The current ladder-bucket method is inefficient as workers have to conduct a lot of activities that are irrelevant to apple picking, such a moving/climbing/descending ladders and transporting apples into a bin. Additionally, apple harvest would be prone to causing occupational injuries (e.g., sprain and strain). More serious scenario occurs when a worker falls from the ladder, which may lead to fracture or even death [5]. The U.S. apple industry faces a challenge that there is insufficient labor to complete the harvest. The insufficient labor is because of the shrinking labor pool, and is worsen by the fact that the manual harvest requires high physical strength. The worker needs to wear a bucket during the entire harvest process to hold apples, and when full, the bucket weighs about 20 kg. Not each employee is capable of such work, and this worsens the shortage for harvest labor [6]. Mechanical harvest of fresh market apples is an approach to address the shortcomings associated with the current ladder-bucket method.

During the past decades, researchers and practical engineers have invested tremendous efforts to realize the mechanical harvest of fresh market apples, which can be categorized into bulk harvesting, harvest robots, and harvest platform [6, 7]. The bulk harvesting represents multiple apples detached from the tree simultaneously, which is usually realized by holding and shaking the tree trunks/limbs. Bulk harvesting is the first mechanical harvest method tested by researchers, and based on this principle, a number of machines have been developed and tested. Since causing excessive bruises during harvest, the developed bulk harvesting machines were not commercially used [8, 9]. On the opposite side of bulking harvesting, instead of detaching multiple apples at a time, the harvest robot picks one apple at a time. To realize individual apple's picking, the harvest robot consists at least two components: machine vision and end-effector. The machine vision system is responsible to identify and localize apples, which information would be used to guide an end-effector to pick the target apple. Several apple harvest robot prototypes were developed and tested, but due to inefficiency, high-cost, and unreliability for long-term use, they are still far away from practical applications [1, 10]. Since both bulk harvesting and harvest robots could not meet the requirements of practical applications, starting from late 1990s, researchers started to explore the harvest platform concept to assist workers in apple harvest [11]. The harvest platform concept provides a platform to replace the role of ladders, and workers stand on the platform to complete the harvest activities. Major advantages of the harvest platform include, but are not limited to, high efficiency, low strength requirements, and reduced occupational injuries [12, 13].

Though a number of research projects have been conducted during the past decades on mechanical harvest of fresh market apples with promising outcomes published [14–18], and several review papers have been published in machine vision for apple detection and localization [19, 20] and bulk harvesting [6], no reviews were found about the mechanical harvest of fresh market apples in a holistic and systematic manner. Thus, this chapter aims to fill this gap by reviewing the technology evolvement in terms of bulk harvesting, harvest robot, and harvest platform.

1.2 Bulk Harvesting

The bulk harvesting method is to detach multiple apples from the tree at one time, and a catch device is placed below the apples for catching purposes [21]. According to the different approaches used for apple detachment, bulk harvesting could be grouped into four categories: shake-and-catch, combing, rod press, and air jet.

1.2.1 Shake-And-Catch Method

The shake-and-catch method generates inertial force for the apples, and when the inertial force applied to apples is larger than the bonding force between apples and limbs, fruit are detached automatically. Since the inertial force is applied to a limb or a branch (not a specific apple) and there are a lot of apples on a limb or branch, multiple apples would be detached simultaneously [21].

Researchers from the Cornell University developed a shake-and-catch principle fresh market apple harvester (Fig. 1.1; [22]). The machine would hold the tree trunk

Fig. 1.1 Cornell apple harvester with extended catching surfaces to be inserted into the canopy for apple collection and transportaion

and then shake. A design advantage of the system is that it was mounted with a set of three collection surfaces, instead of one surface, which can significantly decrease the apple free-fall distance after detachment. By significantly reducing the apple free-fall dropping distance, it would lower the bruising rate. Additionally, the top and middle collection surfaces are inclined, in which manner the gravity would transport apples automatically to a main conveyor. For the bottom horizontal surface, it was powered by motors, instead of gravity, for apple transportation, which also take apples to the main conveyor. The main conveyor would fill apples into a bin. Extensive field tests of the machine on different apple cultivars (e.g., 'McIntosh' and 'Golden Delicious') indicated that the machine was incapable for fresh market apple harvest due to excessive bruising and fruit cuts/punctures, and detailed studies demonstrated that the bruising incurred in many steps. The bruising even incurred before apples were detached from the limb because of apple-to-apple collisions during the detachment. Though the apple free-fall distance was shortened due to the multi-layer catching surfaces compared to the one-layer design, the bruising still incurred because apple-to-limb collisions, as well as the latter arrived apples hit the formally arrived ones at the collection surface.

The above research focused on the traditional tall apple trees with large canopies; however, there are other training systems (tree architectures) such as the dwarfed trees, which are smaller and less complex. Allshouse and Morrow [8] developed an over-the-tree harvester for the dwarfed apple trees (Fig. 1.2). A key component of the harvester is a series of shaking rods that are inserted into the tree canopy for apple detachment. After apples fall onto the bottom collection surface, they were transported into the bins via conveyors. The study was conducted for multiple years for system improvement. The first year's field experiment indicated that 90% of apples were successfully detached from the limb, and it took about 1 min to complete the harvest of one tree. However, the bruising conditions were unsatisfactory with >90% of the harvested apples being bruised. Thus, the harvested apples could not

(a) (b) (c)

Fig. 1.2 Front **a** and back **b** view of a harvester designed for the dwarfed apple trees, and the harvester in working mode **c**

meet the requirements for fresh market consumption. The shaking rods, catching surface and shaker were then all modified significantly with an aim of reducing the bruising ratio. Field tests on the second year with the modified machine generated a promising result that about 52% of the harvested apples were not bruised. The machine went through another round of modification, and the next year's results showed more promising results—65% of the harvested apples were not bruised. Though the machine performance was improved gradually and significantly during the successive three-year field trials, the final version was still far away from meeting the requirements to harvest fresh market apples due to unacceptable bruising.

In the same time period, researchers from United States Department of Agriculture and Oregon State University [23] jointly developed a straddle-frame fresh market apple harvester, consisting of a mechanism to hold and shake the tree trunk and a catching surface located below the tree canopy (Fig. 1.3a). Field experiments on 'Red Delicious' and 'Golden Delicious' apples showed that 47% and 79% of harvested apples were severely damaged, respectively, which were not suitable for fresh market apple harvest. In addition to bruising, stem pull was also a serious issue, as more than 80% of harvested apples came without the stem, and this would pose a challenge

(a) (b) (c)

Fig. 1.3 Straddle-frame apple harvester: **a** machine front view; **b** harvester filled with spheres; and **c** separating spheres from fruit after harvest

for postharvest handling (e.g., storage). To mitigate bruising conditions and alleviate stem pull issue, the group initiated and tested a plastic sphere concept [24]. The plastic sphere concept was to fill the space between fruit and limbs/branches (Fig. 1.3b), after which the tree was shaken for fruit detachment. The plastic sphere actually performed a cushion to avoid serious collisions between two fruits as well as between fruits and limbs/branches. Low-density polyethylene spheres, with a 75 mm diameter, were tested, and field experimental results demonstrated that they reduced apple bruising ratio by 39% and 10% for 'Red Delicious' and 'Golden Delicious', respectively. A major shortcoming of the plastic sphere concept is that it significantly increased the harvest time. Before the harvest/shaking, the plastic spheres needed to be filled and after the harvest, they had to be separated from the real apples (Fig. 1.3c). Efficiency analysis showed that for harvesting a tree, applying the plastic spheres would cost more than 13 times of time compared to not using them, and economic evaluation concluded that the reduction in apple bruising could not offset the increased time [25].

All the above described harvesters work in the stop-and-go mode: the machine would stop at a tree and then harvest apples of the tree, after which the machine would move to the next tree for harvesting. Since the stop-and-go approach significantly lowers the overall efficiency, Monroe [26] and Peterson [27], to improve the efficiency, developed an over-the-row continuous apple harvester (Fig. 1.4a). The machine is extendable for easy transportation, with the narrow form shown in Fig. 1.4b. A crucial engineering contribution of this study is the development of a new shaker (Fig. 1.4c) that allows the shaking and machine forward moving occurring simultaneously. The shaker is attached to the machine frame with a track, which allows the relative movement between the shaker and track (machine). In the working mode, the machine moves forward (so the track), while the shaker holds the tree trunk to complete a certain amount of shaking time (e.g., 20 s). After the shaking amount of time lapses, the shaker would be removed out of the trunk, and powered by a motor to move forward (quickly than the machine moving speed) to the next tree trunk. Extensive experimental results showed that the machine could complete one tree harvest within 15 s. In addition to apple harvest, the machine could also be used for pruning, spraying, and thinning. Though the machine could harvest apples

Fig. 1.4 **a** Over-the-row continuous apple harvester in working mode; **b** apple harvester in narrow mode for easy transportation; and **c** new shaker design allowing machine travelling during shaking

Fig. 1.5 Straddle-type continuous apple harvester: **a** front and **b** back view

efficiently on a continuous mode, the unsatisfactory fruit bruising conditions made it infeasible to harvest fresh market apples.

Peterson et al. continued to work on the continuous principle and constructed another straddle-type apple harvester (Fig. 1.5). Compared to the preliminary version of the continuous harvester (Fig. 1.4), this machine is more advanced, consisting of an impact shaker, a catching surface, and conveyors. Field experiments demonstrated that 'Extra Fancy' + 'Fancy' grade (Table 1.1) apples were 70, 81, 79, and 51% for 'Stayman', 'York', 'Delicious', and 'Golden Delicious', respectively. Though the group continued with the machine improvements in terms of shaker, conveyor, and apple collection surfaces, the system still could not meet the requirement of harvesting fresh market level apples [9, 28]. To be capable of harvesting fresh market apples, it requires at least 95% of the fruit that have been handled by the system would be in Extra Fancy level [29].

In addition to the above mentioned apple harvesters based on the shake-and-catch concept, several others machines were developed by Peterson and Monroe ([31], Fig. 1.6a), McHugh et al. ([32], Fig. 1.6b), and Peterson and Wolford ([33], Fig. 1.6c). A common point of these harvesters is that they have a shaker and a large catching surface. All these machines work efficiently, with a harvest speed ranging from 60 to 240 trees/h. Due to the excessive bruising issue, these machines were

Table 1.1 Classification of apple bruise damage [30]

Class	USDA fresh market standards	Bruise specification
1	"Extra Fancy"	No bruising
2	"Extra Fancy"	Bruise diameter ≤ 3.2 mm (1/8 in.)
3	"Extra Fancy"	Bruise diameter 3.2 to 6.4 mm (1/ 8 to 1/4 in.)
4	"Extra Fancy"	Bruise diameter 6.4 mm (1/4 in.) to 12.7 mm (1/2 in.) or area of several bruises < 127 mm^2
5	"Fancy"	Bruise diameter 12.7 to 19 mm (1/2 to 3/4 in.) or total area of multiple bruises < 283 mm^2
6	Downgraded	Bruises larger than the tolerances in "Fancy"
7	Downgraded	Cuts or punctures of any size

Fig. 1.6 Different apple harvesters based shake-and-catch concept

capable for harvesting processing apples, but challenging to be used for harvesting fresh market apples.

1.2.2 Combing Approach

Since all the developed harvesters based on the shake-and-catch concept could not generate satisfactory results in apple bruising, researchers attempted a combing principle, which consists of a multiple rods mechanism similar to a comb [34]. The combing mechanism should meet specific design criteria: (1) soft, to minimize apple bruising; (2) rigid, to be capable of detaching apples; and (3) resilient, to return to their original positions after deflection. Since the gap between two neighboring rods is smaller than the diameter of an apple, the mechanism would remove apples while moving forward. Le Flufy [35, 36] developed an apple harvester based on the combing concept, and field experiment showed that the system successfully detached 90% of 'Cox' apples from trees. Unfortunately, further studies were not followed. There are still a lot of work pending before the machine comes to the commercial stage, such as developing a high-throughput apple collection mechanism and conducting economic evaluation (Fig. 1.7).

Fig. 1.7 An apple harvester based on combing concept in field test

Fig. 1.8 An apple harvester based on rod press concept in field test

1.2.3 Rod Press Principle

Another apple detaching approach is rod press, which is based on a rod array mechanism. While the rod array mechanism moves downwards, the rod soft end section would push apples downwards slowly. With the rods keep on moving downwards, the force applies to the apples increases, and when the force is larger than the bonding force between apples and limbs, apples are automatically detached. Based on the rod press principle, an apple harvester prototype was developed and tested (Fig. 1.8; [37]). The machine was tested on T-trellis orchards with about 90% of harvested apples categorized into 'Extra Fancy' and 'Fancy' [9, 38]. A key limitation for this machine to be widely applied is that it is only tested on the T-trellis tree canopies, with its performance on other types of tree canopies unknown. The T-trellis tree canopy is not popularly adopted.

1.2.4 Air Jet Mechanism

In addition to the shake-and-catch, combing, and rod press, air jet is another approach developed to detach apples. The air jet principle takes advantage of opposite jets of air across the tree canopy to generate apple removal force. Since the air jet force is relatively small compared to the other three approaches, abscission chemicals (e.g., ethephon) are usually used to reduce the bonding force between apples and limbs. Figure 1.9 shows a constructed apple harvester based on this concept [39, 40]. Field experiments demonstrated that proper timing between the application of ethephon and air jet harvester application is challenging—if the time window is too

(a) (b)

Fig. 1.9 Side (**a**) and back view (**b**) of the air jet apple harvester

short, apples could not be detached by the air jets, if the time window is too long, apple already fall off the tree before the machine starts to harvest. Additionally, the bruising conditions of the harvested apples are severe due to collision between apples and limbs/braches and between different apples, as the air jets increase the collision magnitude. Thus, the air jet based apple harvester could not be used to harvest fresh market apples.

1.3 Harvest Robot

For the bulk harvesting, workers are still involved with some activities (e.g., driving the machine), while for the apple harvest robots, the workers are not involved, as the whole harvest process is fully automatic. Compared to the robots that have been used widely in a number of industries (e.g., car assembly line), application of robots in agriculture is way more behind, due to the complex field working conditions (e.g., uneven terrain, vibration, and dust) [19, 41–43]. During the past decades, tremendous efforts on developing apple harvest robots have been invested, and progress has been made. In general, an apple harvest robot consists of three major work: apple detection, apple localization, and end effectors [1, 44].

1.3.1 Apple Detection

A harvest robot needs to first identify individual apples from the noisy background, including leaves, limbs, branches, and apple overlapping. Researchers usually use image segmentation, followed by machine learning (ML) algorithms to detect apples. Commonly used ML algorithms include support vector machine (SVM), neural network (NN), random forest, and k-nearest neighbors [45, 46]. Very recently, with rapid progress in deep learning (DL), researchers start to take advantage of different DL algorithms on apple detection [1].

Wang et al. [47] tested ML algorithms on detecting apples from noisy background. The collected images were first segmented using region growing method and color properties. After features of color and shape were extracted, they were fed into SVM and NN. Experimental results showed that the SVM had a more satisfactory performance (89% accuracy) over the NN (67% accuracy). After incorporating texture features, the model resulted in a 90% accuracy in detecting 'Golden Delicious' and 'Jonagold' apples [48, 49].

Very recently, researchers start to test and validate innovative models (especially new DL algorithms) on fruit detection [50–52]. Gao et al. [53] applied Faster R-CNN model on detecting apples under different conditions: non-, leaf-, branch/wire-, and fruit-occlusion. After augmenting the originally collected 800 images from orchards, a 12,800 image dataset was obtained. The Faster R-CNN model generated an average accuracy of 88%, with 0.2 s needed to process an image. Thus, the developed methodology could be used to guide the apple harvest robot for real-time apple detection and picking. Gené-Mola et al. [54] and Wan and Goudos [55] optimized the convolutional and pooling layers of the Faster R-CNN model to improve the accuracy and reduce computation cost, and the improved model achieved an accuracy of 95%. In addition to the Faster R-CNN, a number of other DL algorithms, such as Yolov3 and VGG16 were also tested for the purpose of apple detection, and experimental results showed that the accuracy was above 95%, and individual image's processing time was within 1 s. Thus, DL models (e.g., Yolov3) could meet the requirements in terms of accuracy and processing time for real-time apple detection.

In addition to color images, researchers explored the performance of other types of images (e.g., thermal and multi-spectral) on apple detection [44, 56]. Since apples and objects performing as noisy background (e.g., limb and leaf) may have different temperature, [57] collected thermal images for apple detection. By taking and analyzing thermal images in the afternoon, the correlation coefficient between the number of apples detected in the thermal images and ground truth data was 0.85. Furthermore, [58] fused thermal and color images to improve fruit detection accuracy, and demonstrated the fusion approach generated a higher accuracy over using either type of the images. Safren et al. [59] took advantage of hyperspectral images for green apple detection. Using principle component analysis, researchers first reduced the high dimension data from 41 to 15 bands. After applying machine vision techniques (e.g., blob analysis), it achieved a 90% accuracy.

In addition to the technology progress in engineering, apple tree canopies are evolving from large, complex, and unstructured pattern to small, simple, and structured format [1]. Figure 1.10a shows the conventional apple tree canopy—though it is large and complex, it is still the main type for old orchards in many states, such as Michigan and Pennsylvania states. Figure 1.10b and c are the new apple tree canopy type, which is the dominant format in Washington state. In addition, through the U.S., when new orchards are established, a majority of them use the new tree canopy. Since the new canopy is simple, small, and structured, it facilitates the detection of apples.

(a) (b) (c)

Fig. 1.10 Apple tree canopy evolvement: **a** Conventional large, complex, and unstructured canopy, **b** new tall spindle structure, and **c** V-Trellis structure

1.3.2 Apple Localization

After detecting apples from trees, the apple localization information is needed to drive/guide the end-effector to pick apples accurately. Inaccurate apple localization information would either result in failure grasping or bruising of apples. During the past decades, researchers have conducted a myriad of studies on accurately locating apples [1].

In the early stage studies, researchers usually mounted the camera at the center of the end-effector, in which way it only needed to adjust the end-effector position to keep apples at the center of the collected images [60–62]. Though this method is validated, it is overall inefficient because it only focuses on one specific apple. Since there are a lot of apples in the tree, a more reasonable and efficient approach is to localize multiple apples simultaneously, and then make a plan on the picking route to reduce the total harvesting time (higher efficiency). Stereovision, consisting of multiple physically separated cameras, was then applied to localize multiple apples in the tree [1]. A major challenge while applying stereovision is to find the reference points in different images. The reference point refers to a point in the physical world that shows up in different images, and the identification of reference points is named as corresponding problem [63]. Many researchers have invested efforts to address this corresponding problem by either narrowing down the images or using one detected apple as a reference, and field experimental results showed that the stereovision systems generated an error within $\pm$ 20 mm [64, 65].

Beyond stereovision, the newly released red–green–blue-depth (RGB-D) camera, which can provide both color and distance information, is another option for apple localization. Zheng et al. [66] and Lin et al. [67] used a RGB-D camera (Microsoft Corporation, Redmond, WA, U.S.) for apple localization, and the errors in the x, y, and z directions were 7 ± 2 mm, -4 ± 3 mm, and 13 ± 3 mm, respectively. The results indicate that the RGB-D camera is a promising candidate to localize apples. After apples are localized, it needs to determine the picking route (Fig. 1.11) with the boundary condition of minimal total picking time. Researchers usually treat the route optimization question as a travelling salesman problem (TSP) to calculate the optimal solution [68].

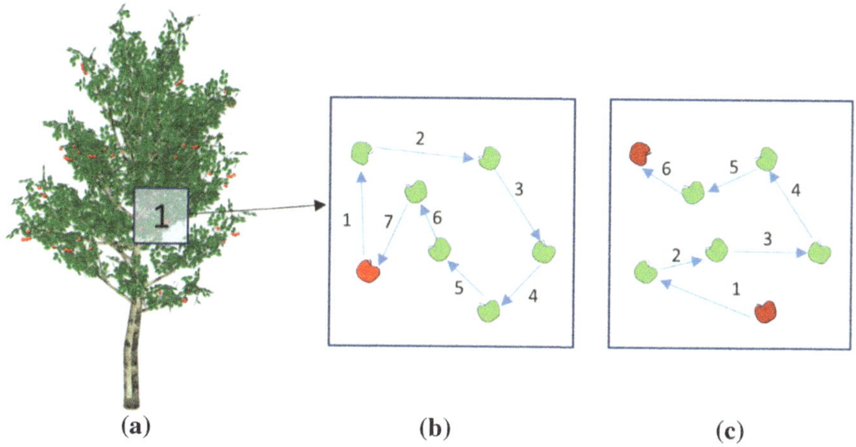

Fig. 1.11 Apple picking route determination: **a** camera view, **b** one potential picking route, and **c** another potential picking route

1.3.3 End-Effector

After apples are detected, localized, and picking route determined, the next step is to drive end-effector to pick apples [69]. There are many end effectors that have been developed and tested over the past decades. Baeten et al. [70] developed a vacuum-based funnel-shape apple gripper (Fig. 1.12). When the end-effector moves close to the target fruit, it would suck and then hold the apple, and when the end-effector retracts, the apple is detached. Instead of using vacuum, [71] developed a spoon-shaped end-effector (Fig. 1.12). Since the spoon shape is complementary to the apple surface, this design could hold apples firmly. Simulating human hands/fingers, a research group at Washington State University developed a three-finger end-effector to detach apples (Fig. 1.12c), after which they continued with the study, and developed an updated version—the soft finger end effector (Fig. 1.12d, [72–74]). All these

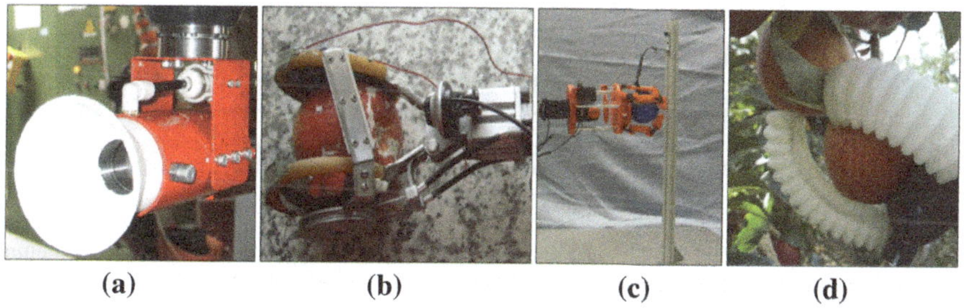

Fig. 1.12 Four different types of end-effectors: **a** vacuum suction, **b** spoon-shaped, **c** three-finger, and **d** soft fingers

designs were tested extensively, but finally they could not make through the commercial application for a number of reasons: (1) low efficiency, (2) apple bruising, and (3) limited stroke (cannot get access to apples in the canopy), and (4) cannot harvest apples grown in cluster.

1.3.4 Integrated Apple Harvest Robots

Beyond the end-effectors described in the above section, there are two more types of end-effectors that are on their way for commercialization. Additionally, based on each end-effector, a harvest robot was integrated.

The vacuum-tube end-effector takes advantage of vacuum to suck apples (Fig. 1.13a). Instead of holding apples using the tube (Fig. 1.12a), this design sucks apples into the tube, and the tube continues to transport apples into a bin. Based on this design, an apple harvest robot was integrated (Abundant Robotics, Inc. CA, U.S.A.; Fig. 1.13c). The robot has been tested multiple years and the company keeps on optimizing the system [1, 75]. The other end-effector consists of a telescoping arm and a finger end-effector (Fig. 1.13b), based on which a complete system was designed and tested (Fig. 1.13d). Compared to the vacuum approach, this design has

Fig. 1.13 Two end-effectors and their associated robots on the way for commercial use: Vacuum-tube end-effector (**a**) and its corresponding robot (**c**); finger end-effector (**b**) and its corresponding robot (**d**)

a telescoping arm, which significantly increases the stroke. Both robots are still in the extensive test stage, but both robots need to fix some bottlenecks before they are commercially applied, such as how to pick apples grown in cluster.

1.4 Harvest Platform

Since the bulk harvesting could not meet the bruising requirements for fresh market apples, and the harvest robots still have a long way ahead before they come to the commercial application, researchers and agricultural engineers started to put efforts since late 1990s on developing harvest platforms to replace the conventional ladder-bucket harvest method [12]. A majority of the shortcomings associated with the traditional ladder-bucket method is associated with the application of ladders, such as ladder moving, and climbing/descending ladders. While using a platform to replace the ladders, the ladder related activities are eliminated. Thus, applying harvest platform should be able to improve the harvest efficiency, reduce strength requirements, and alleviate occupational injuries [76–78].

Peterson and Miller [11] developed a harvest platform, and workers sit on the platform to pick apples (Fig. 1.14a). The seats are height adjustable to provide a convenient picking position for workers. After apples are picked, workers just need to drop the apples, after which a specifically designed fruit catching surface would be responsible to catch and transport apples. Experimental results showed that the fruit quality harvested by using the platform was as good as those harvested by the conventional ladder-bucket method.

A vacuum apple harvest platform (Phil Brown Welding Corp., Conklin, MI, U.S.) was developed and tested by Luo et al. [79]. Instead of dropping apples, workers put apples into a vacuum tube, which is responsible for apple transportation. Due to the

(a) (b)

Fig. 1.14 A platform that workers can sit on to pick apples (**a**) and fruit catching and transporting mechanism (**b**)

high speed during transportation, apples would first go through a speed reduction mechanism before they are delivered into the bin. An impact recording device (IRD, Techmark, Inc., Lansing, MI, U.S.) was used to evaluate the collision magnitude while apples going through the system. Based on the IRD data, it was estimated that 99% of apples harvested by this vacuum platform could meet the 'Extra Fancy' (Table 1.1) requirements. Schupp et al. [12] tested the machine in orchards with real apples, and experimental results showed that more than 92% of the harvested apples ('Honeycrisp', 'Fuji', and 'Golden Delicious') were graded as 'Extra Fancy' and 'Fancy' (Fig. 1.15).

In addition to the vacuum type platform, there are some other harvest platforms available on the market [16]. The Huron system (HuronFruitSystems, Neward, NY, U.S.) provides elevations attached to the platform for workers to harvest high level apples (Fig. 1.16a). For the low level apples, workers stand on the ground for harvesting. The system carries on bins, and when the bags are full for workers, they

Fig. 1.15 Vacuum apple harvest platform

(a) (b) (c)

Fig. 1.16 Different types of harvest platforms on the market: **a** Huron system, **b** precision platform, and **c** pluk-o-track platform

walk to the bin to dump apple gently into the bin. A similar platform (Precise Manufacturing, Inc., Casnovia, MI, U.S.) is developed, but with less bins (three) carried on the machine (Fig. 1.16b). Different from using bags to temporarily hold apples, a pluk-o-track platform is designed, which uses conveyor to carry apples (Fig. 1.16c; Munckhof Fruit Technology Innovators, Venrayseweg, The Netherlands). Working with the pluk-o-track, workers only need to put apples onto the conveyors, and the conveyors would transport apples into the bin, in which manner workers' time on transporting apples into the bin is saved.

Overall, there are several types of harvest platforms having been adopted by apple growers, but the rate is low, due to the machine's high price ($50,000–$120,000) and the economic benefits are uncertain. The adoption rate is especially low for small orchards as they are more concerned on the economic benefits and cannot afford large amount of investment. There are two potential options to address these issues: develop a small and low-cost harvest platform (<$35,000), and add more functions to the platform (e.g., infield sorting) then allow the machine to bring more benefits to apple growers.

1.5 Conclusion

The paper discusses the technology evolvement of mechanical harvest of fresh market apples in terms of bulk harvesting, harvest robots, and harvest platform. For the bulk harvesting, though many prototypes have developed, tested, and improved, all existing machines cause excessive bruising, which make them unsuitable to be used to harvest fresh market apples. For the harvest robots, though the two state-of-the-art prototypes have been tested and improved for multiple years, there are a few bottlenecks to be fixed, such as how to harvest apples grown in cluster. Thus, there is still a long way before their commercial application. The harvest platform is the currently only available technology that has been slowly adopted by apple growers. However, the adoption rate is very low and all the existing harvest platforms are only suitable for large-scale orchards, due to their high price ($50,000~$120,000) and large size. Thus, more efforts should be invested to develop a small and low-cost harvest platform and add more functions onto the platform (e.g., sorting) to have it bring more economic benefits to apple growers.

Credit Authorship Contribution Statement **D. Hu**: Writing—original draft. **Y. Ampatzidis**: Writing—review & editing. **G. Liu**: Writing—review & editing. **Z. Zhang**: Conceptualization, Investigation, Methodology, Project administration, Writing—original draft, Writing—review & editing. **K. Betitame**: Writing—review & editing.

Disclaimer
Mention of commercial products or orchards in this paper is only for providing factual information and does not imply endorsement of them by authors over those not mentioned.

Declaration of Competing Interest
The authors declare that they have no known competing financialinterests or personal relationships that could have appeared to influence the work reported in this paper.

References

1. Zhang Z, Igathinathane C, Li J, Cen H, Lu Y, Flores P (2020) Technology progress in mechanical harvest of fresh market apples. Comput Electron Agricul 175:105606. https://doi.org/10.1016/j.compag.2020.105606
2. Hyun TK, Jang KI (2016) Apple as a source of dietary phytonutrients: an update on the potential health benefits of apple. EXCLI J 15:565
3. Gallardo K, Taylor M, Hinman H (2010) 2009 Cost estimates of establishing and producing gala apples in Washington. Washington State University, Pullman, WA
4. Freivalds A, Park S, Lee C, Earle-Richardson G, Mason C, May JJ (2006) Effect of belt/bucket interface in apple harvesting. Int J Ind Ergon 36(11):1005–1010. https://doi.org/10.1016/j.ergon.2006.08.005
5. Fathallah FA (2010) Musculoskeletal disorders in labor-intensive agriculture. Appl Ergon 41(6):738–743. https://doi.org/10.1016/j.apergo.2010.03.003
6. Zhang Z, Heinemann PH, Liu J, Baugher TA, Schupp JR (2016a) The development of mechanical apple harvesting technology: a review. Trans ASABE 59(5):1165–1180. https://doi.org/10.13031/trans.59.11737
7. Peterson DL, Bennedsen BS, Anger WC, Wolford SD (1999) A systems approach to robotic bulk harvesting of apples. Trans ASAE 42(4):871–876. https://doi.org/10.13031/2013.13266
8. Allshouse GW, Morrow CT (1972) Over-the-row harvester for dwarf fruit trees. Trans ASAE 15(6):1038–1043. https://doi.org/10.13031/2013.38067
9. Peterson DL, Miller SS (1989) Advances in mechanical harvesting of fresh market quality apples. J Agric Eng Res 42(1):43–50. https://doi.org/10.1016/0021-8634(89)90038-3
10. Zhang Z (2015) Design, test, and improvement of a low-cost apple harvest-assist unit. Ph.D. diss. Pennsylvania State University, Department of Agricultural and Biological Engineering, University Park, PA
11. Peterson DL, Miller SS (1996) Apple harvesting concepts for inclined trellised canopies. Appl Eng Agric 12(3):267–271. https://doi.org/10.13031/2013.25648
12. Schupp J, Baugher T, Winzeler E, Schupp M, Messner W (2011) Preliminary results with a vacuum assisted harvest system for apples. Fruit Notes 76(4):1–5
13. Zhang Z, Heinemann PH, Liu J, Schupp JR, Baugher TA (2016b) Design and field test of a low-cost apple harvest-assist unit. Trans ASABE 59(5):1149–1156. https://doi.org/10.13031/trans.59.11708
14. Bac CW, van Henten EJ, Hemming J, Edan Y (2014) Harvesting robots for high-value crops: State-of-the-art review and challenges ahead. J Field Robot 31(6):888–911. https://doi.org/10.1002/rob.21525
15. He L, Fu H, Karkee M, Zhang Q (2017) Effect of fruit location on apple detachment with mechanical shaking. Biosys Eng 157:63–71. https://doi.org/10.1016/j.biosystemseng.2017.02.009
16. Lu R, Zhang Z, Pothula AK (2017) Innovative technology for apple harvest and in-field sorting. Fruit Q 25(2):11–14
17. Pothula AK, Zhang Z, Lu R (2018) Design features and bruise evaluation of an apple harvest and in-field presorting machine. Trans ASABE 61(3):1135–1144. https://doi.org/10.13031/trans.12327
18. Zhang Q, Karkee M (2016) Fully automated tree fruit harvesting. Resour Mag 23(6):16–17

19. Gongal A, Amatya S, Karkee M, Zhang Q, Lewis K (2015) Sensors and systems for fruit detection and localization: a review. Comput Electron Agric 116:8–19. https://doi.org/10.1016/j.compag.2015.05.021
20. Zhao Y, Gong L, Huang Y, Liu C (2016) A review of key techniques of vision-based control for harvesting robot. Comput Electron Agric 127:311–323. https://doi.org/10.1016/j.compag.2016.06.022
21. Peterson DL (2005) Harvest mechanization progress and prospects for fresh market quality deciduous tree fruits. HortTechnology 15(1):72–75. https://doi.org/10.21273/horttech.15.1.0072
22. Millier WF, Rehkugler GE, Pellerin RA, Throop JA, Bradley RB (1973) Tree fruit harvester with insertable multilevel catching system. Trans ASAE 16(5):844–850. https://doi.org/10.13031/2013.37641
23. Berlage AG, Langmo RD (1974) Harvesting apples with straddle-frame trunk shaker. Trans ASAE 17(2):230–232. https://doi.org/10.13031/2013.36829
24. Berlage AG, Langmo RD (1976) Trunk shaker harvesting of apples surrounded by plastic spheres. Trans ASAE 19(2):209–0212. https://doi.org/10.13031/2013.35996
25. Nivon M, Sarig Y, Tenne Y, Hial D, Shnaps B, Manor A, Emek M (1977) Mechanical harvesting of apples in a damping medium. Trans ASAE 20(3):457–460. https://doi.org/10.13031/2013.35577
26. Monroe GE (1982) An over-the-row continuous tree-crop harvester. Trans ASAE 25(4):888–892. https://doi.org/10.13031/2013.33634
27. Peterson DL (1982a) Continuously moving over-the-row harvester for tree crops. Trans ASAE 25(6):1478–1483. https://doi.org/10.13031/2013.33749
28. Peterson DL, Miller SS, Whitney JD (1994) Harvesting semidwarf freestanding apple trees with an over-the-row mechanical harvester. J Am Soc Hortic Sci 119(6):1114–1120. https://doi.org/10.21273/jashs.119.6.1114
29. Zhang Z, Pothula AK, Lu R (2017) Development and preliminary evaluation of a new bin filler for apple harvesting and in-field sorting machine. Trans ASABE 60(6):1839–1849. https://doi.org/10.13031/trans.12488
30. Peterson DL, Tabb AL, Baugher T, Lewis K, Glenn DM (2010) Dry bin filler for apples. Appl Eng Agricul 26(4):541–549. https://doi.org/10.13031/2013.32057
31. Peterson DL, Monroe GE (1977) Continuously moving shake-catch harvester for tree crops. Trans ASAE 20(2):202–205. https://doi.org/10.13031/2013.35523
32. McHugh CM, Webb BK, Hood CE, Garrett TR (1981) A compact, single-operator tree crop harvester. Trans ASAE 24(1):26–30. https://doi.org/10.13031/2013.34191
33. Peterson DL, Wolford SD (2003) Fresh–market quality tree fruit harvester part II: apples. Appl Eng Agric 19(5):545. https://doi.org/10.13031/2013.15314
34. Le Flufy MJ (1982a) The design of a prototype apple harvester. J Agric Eng Res 27(1):51–60. https://doi.org/10.1016/0021-8634(82)90057-9
35. Le Flufy MJ (1982b) Harvest trials with a prototype apple harvester. J Agric Eng Res 27(5):415–420. https://doi.org/10.1016/0021-8634(82)90079-8
36. Le Flufy MJ (1983) Apple harvesting by a combing technique. Trans ASAE 26(3):661–664. https://doi.org/10.13031/2013.33997
37. Peterson DL (1982b) Rod press fruit removal mechanism. Trans ASAE 25(5):1185–1188. https://doi.org/10.13031/2013.33694
38. Peterson DL, Kornecki TS (1987) Mechanical apple harvester for T-trellis canopies. Trans ASAE 30(3):597–600. https://doi.org/10.13031/2013.30445
39. Berlage AG (1973) Apple harvesting trials with oscillating air jets. Trans ASAE 16(3):460–461. https://doi.org/10.13031/2013.37542
40. Thomas RL (1964) An engineering investigation in the use of a pulsating air stream to mechanically detach apples from the tree. MS Thesis. Rutgers University, New Brunswick, NJ
41. Ceccarelli M, Figliolini G, Ottaviano E, Mata AS, Criado EJ (2000) Designing a robotic gripper for harvesting horticulture products. Robotica 18(1):105–111. https://doi.org/10.1017/s026357479900226x

42. Sarig Y (1993) Robotics of fruit harvesting: a state-of-the-art review. J Agric Eng Res 54(4):265–280. https://doi.org/10.1006/jaer.1993.1020
43. Sivaraman B, Burks TF (2006) Geometric performance indices for analysis and synthesis of manipulators for robotic harvesting. Trans ASABE 49(5):1589–1597. https://doi.org/10.13031/2013.22033
44. Silwal A, Karkee M, Zhang Q (2016) A hierarchical approach to apple identification for robotic harvesting. Trans ASABE 59(5):1079–1086. https://doi.org/10.13031/aim.20152167504
45. Flores P, Zhang Z, Igathinathane C, Jithin M, Naik D, Stenger J, Ransom, Kiran R (2021) Distinguishing seedling volunteer corn from soybean through greenhouse color, color-infrared, and fused images using machine and deep learning. Ind Crops Prod 161:113223. https://doi.org/10.1016/j.indcrop.2020.113223
46. Jahan N, Flores P, Liu Z, Friskop A, Mathew J, Zhang Z (2020) Detecting and distinguishing wheat diseases using image processing and machine learning algorithms. ASABE Paper No. 2000372. ASABE, St. Joseph, MI. https://doi.org/10.13031/aim.202000372
47. Wang JJ, Zhao DA, Ji W, Tu JJ, Zhang Y (2009) Application of support vector machine to apple recognition using in apple harvesting robot. Int Conf Inf Autom 1110–1115. https://doi.org/10.1109/icinfa.2009.5205083
48. Kurtulmus F, Lee WS, Vardar A (2011) Green citrus detection using 'eigenfruit', color and circular Gabor texture features under natural outdoor conditions. Comput Electron Agric 78(2):140–149. https://doi.org/10.1016/j.compag.2011.07.001
49. Stajnko D, Rakun J, Blanke M (2009) Modelling apple fruit yield using image analysis for fruit colour, shape and texture. Eur J Hortic Sci 74(6):260
50. Fu L, Majeed Y, Zhang X, Karkee M, Zhang Q (2020) Faster R-CNN–based apple detection in dense-foliage fruiting-wall trees using RGB and depth features for robotic harvesting. Biosys Eng 197:245–256. https://doi.org/10.1016/j.biosystemseng.2020.07.007
51. Fu L, Tola E, Al-Mallahi A, Li R, Cui Y (2019) A novel image processing algorithm to separate linearly clustered kiwifruits. Biosys Eng 183:184–195. https://doi.org/10.1016/j.biosystemseng.2019.04.024
52. Majeed Y, Zhang J, Zhang X, Fu L, Karkee M, Zhang Q, Whiting MD (2020) Deep learning based segmentation for automated training of apple trees on trellis wires. Comput Electron Agric 170:105277. https://doi.org/10.1016/j.compag.2020.105277
53. Gao F, Fu L, Zhang X, Majeed Y, Li R, Karkee M, Zhang Q (2020) Multi-class fruit-on-plant detection for apple in SNAP system using Faster R-CNN. Comput Electron Agric 176:105634. https://doi.org/10.1016/j.compag.2020.105634
54. Gené-Mola J, Vilaplana V, Rosell-Polo JR, Morros JR, Ruiz-Hidalgo J, Gregorio E (2019) Multi-modal deep learning for Fuji apple detection using RGB-D cameras and their radiometric capabilities. Comput Electron Agric 162:689–698. https://doi.org/10.1016/j.compag.2019.05.016
55. Wan S, Goudos S (2020) Faster R-CNN for multi-class fruit detection using a robotic vision system. Comp Netw 168:107036. https://doi.org/10.1016/j.comnet.2019.107036
56. Kane KE, Lee WS (2006) Spectral sensing of different citrus varieties for precision agriculture. ASAE Paper No. 061065, ASABE, St. Joseph, MI. https://doi.org/10.13031/2013.21091
57. Stajnko D, Lakota M, Hočevar M (2004) Estimation of number and diameter of apple fruits in an orchard during the growing season by thermal imaging. Comput Electron Agric 42(1):31–42. https://doi.org/10.1016/s0168-1699(03)00086-3
58. Bulanon DM, Burks TF, Alchanatis V (2009) Image fusion of visible and thermal images for fruit detection. Biosys Eng 103(1):12–22. https://doi.org/10.1016/j.biosystemseng.2009.02.009
59. Safren O, Alchanatis V, Ostrovsky V, Levi O (2007) Detection of green apples in hyperspectral images of apple-tree foliage using machine vision. Trans ASABE 50(6):2303–2313. https://doi.org/10.13031/2013.24083
60. Parrish EA, Goksel AK (1977) Pictorial pattern recognition applied to fruit harvesting. Trans ASAE 20(5):822–827. https://doi.org/10.13031/2013.35657

61. Slaughter DC, Harrell RC (1987) Color vision in robotic fruit harvesting. Trans ASAE 30(4):1144–1148. https://doi.org/10.13031/2013.30534
62. Slaughter DC, Harrell RC (1989) Discriminating fruit for robotic harvest using color in natural outdoor scenes. Trans ASAE 32(2):757–0763. https://doi.org/10.13031/2013.31066
63. Wang Q, Nuske S, Bergerman M, Singh S (2013) Automated crop yield estimation for apple orchards. In: Experimental robotics. Springer, Heidelberg, pp 745–758. https://doi.org/10.1007/978-3-319-00065-7_50
64. Si Y, Liu G, Feng J (2015) Location of apples in trees using stereoscopic vision. Comput Electron Agric 112:68–74. https://doi.org/10.1016/j.compag.2015.01.010
65. Takahashi T, Zhang S, Fukuchi H (2002) Measurement of 3-D locations of fruit by binocular stereo vision for apple harvesting in an orchard. ASAE Paper No. 021102, ASABE, St. Joseph, MI. https://doi.org/10.13031/2013.9685
66. Zheng L, Ji R, Liao W, Li M (2018) A positioning method for apple fruits based on image processing and information fusion. IFAC-PapersOnLine 51(17):764–769. https://doi.org/10.13031/aim.20152190688
67. Lin G, Tang Y, Zou X, Xiong J, Fang Y (2020) Color-, depth-, and shape-based 3D fruit detection. Precision Agric 21(1):1–17. https://doi.org/10.1007/s11119-019-09654-w
68. Davidson J (2016a) Mechanical design and field evaluation of a robotic apple harvester. Washington State University, Ph.D. Dissertation
69. Jimenez AR, Ceres R, Pons JL (2000) A survey of computer vision methods for locating fruit on trees. Trans ASAE 43(6):1911–1920. https://doi.org/10.13031/2013.3096
70. Baeten J, Donné K, Boedrij S, Beckers W, Claesen E (2008) Autonomous fruit picking machine: a robotic apple harvester. In Field and service robotics. Springer, Berlin, pp 531–539. https://doi.org/10.1007/978-3-540-75404-6_51
71. Zhao D, Lv J, Ji W, Ying Z, Yu C (2011) Design and control of an apple harvesting robot. Biosys Eng 110(2):112–122. https://doi.org/10.1016/j.biosystemseng.2011.07.005
72. Davidson JR, Mo C (2015) Mechanical design and initial performance testing of an apple-picking end-effector. ASME Int Mech Eng Congr Exposition. Am Soc Mech Eng. https://doi.org/10.1115/imece2015-50482
73. Davidson JR, Silwal A, Hohimer CJ, Karkee M, Mo C, Zhang Q (2016b) Proof-of-concept of a robotic apple harvester. IEEE/RSJ Int Conf Intell Robots Syst (IROS) 634–639. https://doi.org/10.1109/iros.2016.7759119
74. Hohimer CJ, Wang H, Bhusal S, Miller J, Mo C, Karkee M (2019) Design and field evaluation of a robotic apple harvesting system with a 3D-printed soft-robotic end-effector. Trans ASABE 62(2):405–414. https://doi.org/10.13031/trans.12986
75. Wheat D (2019) Robotic apple pickers make progress. Retrieved from https://www.capitalpress.com/ag_sectors/orchards_nuts_vines/robotic-apple-pickers-make-progress/_b1bd180e-1163-11ea-93ec-bf5cfc2aecf1.html
76. Zhang Z, Wang Y, Zhang Z, Li D, Wu Z, Bai R, Meng G (2019a) Ergonomic and efficiency analysis of conventional apple harvest process. Int J Agric Biol Eng 12(2):210–217
77. Zhang Z, Zhang Z, Wang W, Liu H, Sun Z (2019c) The role of a new harvest platform in alleviation of apple workers' occupational injuries during harvest. J Agric Saf Health 25(1):11–24. https://doi.org/10.13031/jash.13103
78. Zhang Z, Zhang Z, Wang X, Liu H, Wang Y, Wang W (2019b) Models for economic evaluation of multi-purpose apple harvest platform and software development. Int J Agric Biol Eng 12(1):74–83
79. Luo R, Lewis KM, Zhang Q, Wang S (2012) Assessment of bruise damage by vacuum apple harvester using an impact recording device. ASABE Paper No. 121338094, ASABE, St. Joseph, MI. https://doi.org/10.13031/2013.41870

Chapter 2
Design, Test, and Improvement of a Low-Cost Fresh Market Apple Harvest-Assist Unit

Y. Shi, Y. Wang, and Zhao Zhang

Abstract The ladder-bucket apple harvest method has a number of shortcomings, including inefficiency, high cost, and causing occupational injuries. Harvest-assist platform is a potential approach to replace the conventional ladder-bucket method. The platform replaces the ladders, and thus could improve the harvest efficiency and avoid ladder fall accidents. Since all existing harvest-assist platforms are large in size, pricy and unsuitable for small orchards, a low-cost apple harvest-assist unit for small orchards was designed, fabricated and tested. The unit consists of a commercially available ORSI platform, and harvest-assist device (receiver, tube, manifold, and distributor) independently designed/fabricated by our team. A distance sensor was used to monitor the apple height in the bin, which can trigger a motor to gradually lift up the harvest-assist device while apples accumulating. The first year field tests demonstrated the multi-hole distributor was the major source for bruising, and based on the first year's field test, a second version cone-shape distributor was designed, which was integrated into the low-cost harvest-assist unit for the second year test. The second year field experiment demonstrated that the optimal distributor speed was 28 revolutions per minute, under which speed the fruit bruising conditions met the fresh market apple requirements. For the high level apples (height > 210 cm from

Y. Shi
College of Engineering, Nanjing Agricultural University, Nanjing 210031, China
e-mail: shiyinyan@njau.edu.cn

Y. Wang
Chinese Academy of Agricultural Engineering, Chinese Society of Agricultural Engineering, Beijing 100125, China
e-mail: wangyingkuan@163.com

Z. Zhang (✉)
Key Lab of Modern Precision Agriculture System Integration Research, Ministry of Education of China, China Agricultural University, Beijing 100083, China
e-mail: zhang.zhao.1@ndsu.edu

Key Lab of Agriculture Information Acquisition Technology, Ministry of Agriculture and Rural Affairs of China, China Agricultural University, Beijing 100083, China

Department of Agricultural and Biosystems Engineering, North Dakota, State University, Fargo, ND 58102, USA

Z. Zhang et al. (eds.), *Mechanical Harvest of Fresh Market Apples*, Smart Agriculture 1,
https://doi.org/10.1007/978-981-16-5316-2_2

the ground) on the tree, the harvest efficiencies were 0.2 and 0.4 apples/s for ladder-bucket method and harvest-assist unit, respectively. This study demonstrated that the newly developed low-cost apple harvest-assist unit could increase 100% harvest efficiency compared to ladder-bucket method with satisfactory bruising conditions on the harvested apples. The developed system provides a potential new approach for harvesting fresh market apples for small orchards.

Keywords Harvest platform · Low cost · Fresh market apples · Bruising · Mechanization · Apple

2.1 Introduction

Apples are one of the most valuable fruit crops in the U.S., which are not only consumed in the nation, but also imported to a huge number of other countries. In addition to bringing billions of dollars to U.S. economy, it provides thousands of jobs [1]. However, in recently years, U.S. apple industry faces a challenge related to labor—insufficient seasonal workers to harvest apples and continuously increasing of labor cost [2–4]. The challenge makes it difficult for U.S. apple industry to beat the competition from other apple production counties (e.g., China and Turkey) [5–8]. To address the labor issue, mechanical harvest of fresh market apples is a potential solution [9, 10].

During the past decades, researchers explored different types of mechanical harvest approaches, which can be categorized into three groups: bulk harvesting, harvest robots, and harvest platform [3]. The bulk harvesting takes advantage of different methods to detach multiple apples from the tree at the same time, such as shaking, pressing, and air jets [11–17]. Though several prototypes based on bulk harvesting concept was developed and tested, they finally did not make into the commercial application for fresh market apple harvest mainly due to excessive bruising. Instead of detaching multiple apples simultaneously, harvest robots can detect and detach apples one at a time [6, 18]. Researchers have integrated a number of apple harvest robots, but all these developed robots are still far from commercial applications, as they all have many bottlenecks to fix (e.g., how to harvest apples grown in cluster).

Since both bulk harvesting and harvest robots could not meet the commercial application requirements for fresh market apple harvest, researchers started to invest efforts to the harvest-assist technologies [3, 19, 20]. The harvest-assist platform technology provides workers a platform to replace the use of ladders. Picking high level apples on the tree by standing on the platform, workers do not need to spend time on moving ladders, as well as climbing and descending ladders. Additionally, by using platforms to replace ladders, ladder fall accidents, which could cause serious results (e.g., fractures and death), can be fully avoided [21]. Based on the harvest-assist concept, a number of harvest platforms are commercially available [22–24]. A vacuum based apple harvest-assist platform (Phil Brown Welding Corp.,

Conklin, MI, U.S.) was developed and tested by Luo et al. [25]. Working with this platform, workers only need to pick and then place apples into the vacuum tube, which is responsible to convey apples to the destination. Experimental results showed that apples harvested by the platform have the same, or even superior quality over apples harvested by the ladder-bucket approach. Instead of using vacuum to transport apples, Munckhof platform (Munckhof Fruit Technology Innovators, Horst, The Netherlands) takes advantage of multiple conveyors to transport apples. Compared to transporting apples automatically (vacuum or conveyor), there are some platforms that still require workers to wear buckets to hold harvested apples temporarily. Since bins are carried on the platform, when the bucket is full, workers only need to walk minimal distance to the bin and then release apples. When the bins are full, all of them would be unloaded at one time, followed by having empty bins loaded. Typical platforms of this concept include Huron (HuronFruitSystems, Neward, NY, U.S.) and Bandits (Automated Ag Inc., Moses Lake, WA, U.S.). All these platforms are commercialized and they all have good performance in terms of improving harvest efficiency and reducing harvest fatigue. In addition, apples harvested by these platforms could meet the fresh market requirements (minimal bruising). All these platforms are pricy, ranging from $50,000 to $120,000 with different configurations [22, 26]. For large apple orchards, commercial harvest platforms are good choices and affordable. However, for small orchards, currently existing platforms are probably too expensive.

To address the high cost issue associated with the existing harvest platforms, it is urgent to develop a low-cost apple harvest-assist unit for the small orchards. Researchers at The Pennsylvania State University initiated the development of a low-cost apple harvest-assist unit targeting small orchards. Since it is aimed for small orchard use, the technology budget was below $35,000, and since it is for harvesting fresh market apples, the quality of harvested apples should be the same as that by using traditional ladder-bucket method. Therefore, the objectives of this study were to: (1) design, test, and improve a low-cost apple harvest-assist unit targeting fresh market apples; (2) compare time percentage spent on apple picking in ladder-bucket vs. harvest-assist method for high level apples; and (3) compare harvest efficiencies of ladder-bucket vs. harvest-assist unit for high level apples.

2.2 Materials and Methods

2.2.1 Integration of the Low-Cost Apple Harvest-Assist Unit

The low-cost apple harvest-assist unit consists of two parts: the mobile platform and the harvest-assist device. To expedite the project progress and device performance, an off-the-shelf ORSI Eco-pick self-propelled mobile platform (ORSI Group, Bologna, Italy) was purchased and used as the base of the unit, on which the harvest-assist device was mounted. The function of the harvest-assist device is to catch harvested apples, and then convey them to the bin without causing bruising damage.

2.2.1.1 Mobile Platform

An ORSI mobile platform was purchased for this project, which can accommodate two workers (Fig. 2.1). The ORSI platform is powered by a set of 12 V DC batteries, and a fully charged set of batteries can last ~6 h. There are two sets of batteries, and while one set is in use, the other battery set is in charge, in which manner the downtime can be avoided. The platform is manually extendable in both horizontal and vertical directions to be adapted to different row configurations and tree heights. The horizontal width is easy to be adjusted as there is a pin that can be assessed by foot: by pressing the pin, the width can be adjusted. The vertical height is a little complex for adjusting, and workers usually take advantage of a jack for the height adjustment, so the height has to be adjusted before or after the harvest, not in the middle of harvest. The platform width ranges from 1.3 to 2.4 m, and the height ranges from 0.9 to 1.6 m [26].

2.2.1.2 Harvest-Assist Device Design

The harvest-assist device was designed and fabricated, consisting of fruit receiver, transportation tube, manifold, distributor, and control system. The fruit receiver is responsible for catching apples picked by workers, after which apples are transported to the manifold via transportation tube. The manifold functions to change the apple movement direction from horizontal to vertical, and simultaneously reduces apple velocity. After exiting from the manifold, apples arrive at the distributor. The distributor first catches apples exiting from the manifold, and then via collisions, apple

Fig. 2.1 ORSI eco-pick mobile platform

velocity is reduced. Then, the spinning distributor carries and then discharges apples into different locations of the bin, after which apples arrive at their final destinations in the bin.

Fruit Receiver

The design of this low-cost apple harvest-assist unit does not require workers to wear a bucket, and thus, there needs a mechanism that receives harvested apples. A receiver was designed and fabricated. The receiver has a tilted bottom, with the end of the slope is an opening. The bottom angle is adjustable, and in this study, it is set as 14° to guarantee each apple can roll automatically downwards. To avoid apple bruising, all the surfaces that have a chance to contact apples are padded with soft foam (weather-and-fire-retardant foam, closed-cell neoprene, McMaster-Carr, Aurora, Ohio, U.S.). The overall dimension of the tube is 30 × 66 × 15 cm (W × L × H), and the large size design provides a large area for workers to drop apples quickly, which can help improve the overall harvest efficiency. The overall design of the receiver is shown in Fig. 2.2.

Fruit Transportation Tube

There is an opening at one end of the receiver, which is connected to a tubing. After apples roll into the opening, the transportation tube would convey apples downwards automatically by gravity. A 153 mm diameter transparent and flexible tube was select. Preliminary lab tests showed that while apples moving through the tube, bruising did not occur [26]. It was also noticed that the tube diameter was large enough to allow apple free movement, without causing apple clogging/stuck.

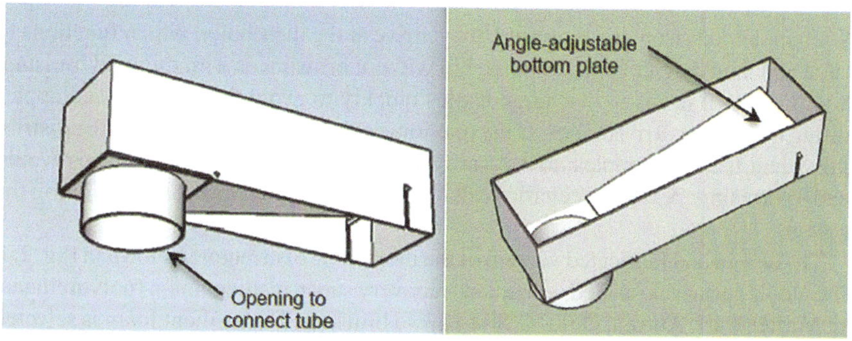

Fig. 2.2 Two views of the receiver design

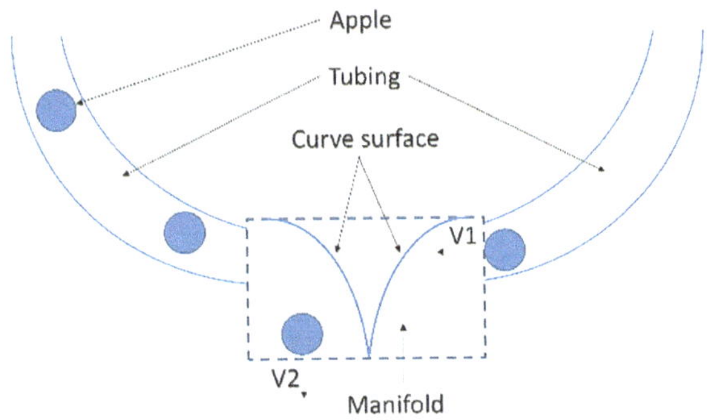

Fig. 2.3 Schematic of manifold to change apple movement direction from horizontal to vertical

Manifold

After getting out of the tube, apples arrive at the manifold, which has two functions as of changing the apple moving direction from horizontal to vertical and reducing the apple velocity (Fig. 2.3). While an apple arriving at the manifold, its direction is horizontal (Fig. 2.3 V1), and after collision with the curve surface, the apple movement direction is changed to be vertically downward (Fig. 2.3 V2). Additionally, when arriving at the end of tube, apples have a high speed as gravitational potential energy is converted to kinetic energy. Since part of the apple kinetic energy is absorbed while collision with curve surfaces, when apples exit from the manifold, their velocities are much smaller than that when entering the manifold. To avoid bruising and absorb more collision energy, the curve surface, as well as any surfaces that have potential to contact apples, is coated with soft foam (weather-and-fire-retardant foam, closed-cell neoprene, McMaster-Carr, Aurora, Ohio, U.S.).

Distributor

After apples exit from the manifold, they arrive at the distributor, which functions to catch and then deliver apples into the bin without bruising or with minimal bruising. Additionally, it needs to discharge apples quickly to avoid the latter arrived apples hitting the former arrived ones. If the previously arrived apples are still at the distributor when the latter apples arrive, serious collisions would occur, and this would lead to bruising. A worse scenario is that it may cause clogging, malfunctioning the system.

The design and fabricated version of the multi-hole distributor is shown in Fig. 2.4. The whole surface is padded with quick-recovery super-resilient foam (polyurethane, McMaster-Carr, Aurora, Ohio, U.S.) to avoid bruising. The resilient foam is selected for its quick recovery characteristic—for the collision area, after the collision with an apple, it needs to be able to recover to the original shape quickly to be ready for the next collision. If the collision area is not recovered to the original shape quickly, when the next apple hits the same area, it would have poor performance on reducing

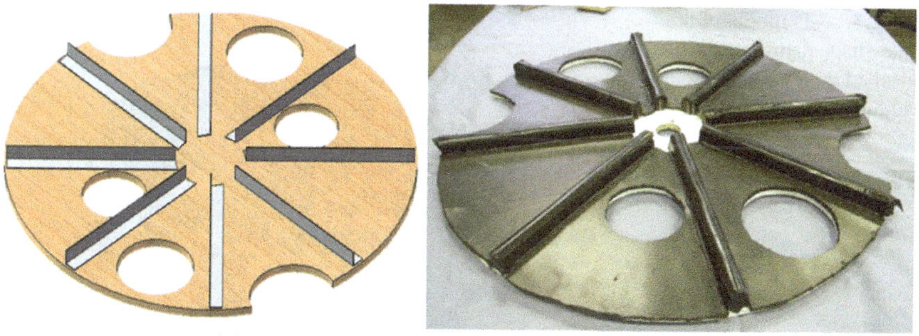

(a) (b)

Fig. 2.4 Multi-hole distributor: **a** schematic and **b** fabricated version

apple speed, which would probably cause bruising damage. The multi-hole design is to help apples have an even distribution in the bin—while the centrifugal force pushes apples from the disc center towards the edge, apples would fall through the holes. Since the distance between the holes and disc center varies, apples would drop off onto spots at different distance to the disc center, which would benefit the fruit uniform distribution in the bin. Since apples exit the distributor with low velocity, it would not cause bruising damage when released into the bin.

The experiment on the multi-hole distributor was conducted (refer to the section of Results and Discussion). Since the multi-hole distributor did not perform satisfactorily, a second version of the distributor was developed. Instead of using holes for the purpose of fruit uniform distribution, the second design was configured with soft padding materials to gently and uniformly release apples into the bin. Compared with the multi-hole design, the soft paddings perform as cushion between apples and bin surfaces to alleviate collision magnitude, as well as reducing bruising. When apples collide with the cone multiple times before they exit, it reduces the apple velocity much more compared to the multi-hole distributor (Fig. 2.5).

Control System

The control system detects the apple height in the bin, and with apple height increases, it would automatically lift up the harvest-assist device. Proper timing of the lifting is critical and too early lifting would generate a large gap between the distributor and apples in the bin. The large gap would lead to a large dropping distance, which could probably result in bruising. If the distributor is lifted too late, when the apples in the bin contact the distributor, it would cause malfunction of the system, and bruise apples significantly.

A photoelectric sensor (FGRW-DT-0A, M.D. Micro Detector, Modena, Italy) were attached to the manifold-distributor, which had a clear view on the apples in the bin. The sensor can detect the distance between distributor and apples in the bin in real time. An empirically determined threshold (30 cm) was pre-set in this study.

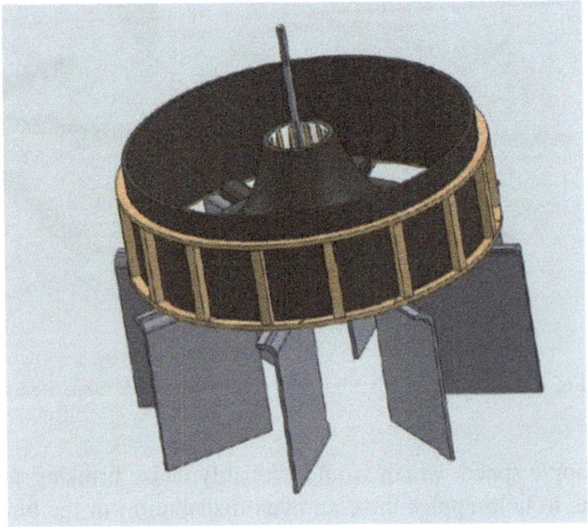

Fig. 2.5. 3-D model of the cone-shape distributor installed with long and soft paddings

The sensor was configured to collect data every 20 s—if the sensor collected distance is smaller than the threshold, it would lift the manifold-distributor 2 cm; otherwise, no action is needed. The distributor spinning speed was controlled by an Arduino with algorithms developed [26, 27]. The speed can be manually adjusted from 6 to 56 revolutions per minute (rpm).

2.2.1.3 Integrated Low-Cost Apple Harvest-Assist Unit

The low-cost apple harvest-assist unit was integrated in the Department of Agricultural and Biological Engineering, The Pennsylvania State University. After preliminary test and multi-rounds improvements, the unit (Fig. 2.6) was transported to the Pennsylvania State Fruit Research and Extension Center, Biglerville, PA, U.S. for field test.

2.2.2 Field Tests

2.2.2.1 First Year Field Test

The first year field experiment aimed at identifying the components causing apple bruising, and find room for improvement. The collected information would be used to improve the machine. This study focused on four components that would potentially cause apple bruising: tube, manifold, distributor, and collection bin. The receiver was not involved in this study as preliminary tests showed it would not bruise apples. To quantify the effects of the tubes on bruising, apples were collected at the end

Fig. 2.6 Integrated low-cost apple harvest-assist unit under field test: **a** original version integrated with multi-hole distributor for the first year test, **b** updated version integrated with cone-shape distributor for the second year test

of tubing, before they arrived at the manifold. Similarly, to understand the performance of manifold, apples were collected immediately when they exited from the manifold, before arriving at the distributor. For the distributor, to avoid the bruising caused by the apple collisions onto bin surfaces, soft padding materials (weather-and-fire-retardant foam, closed-cell neoprene, McMaster-Carr, Aurora, Ohio, U.S.) were used to temporarily cover the bin surfaces, and apples were collected at the bin. Additionally, the effects of bin on bruising were evaluated, and this was realized by removing the padding foams and collecting apples at the bin. During the experiment, the distributor speed was set as 28 rpm. The 28 rpm was determined by preliminary tests—too high speed would bruise apples, and too low speed would lead to apple collisions at the distributor (details see Results and Discussion section).

The experiment was conducted as the Pennsylvania State Fruit Research and Extension Center, Biglerville, PA, U.S. in 2013 harvest season. 'Fuji' variety was used for the experiment. During the experiment, one worker stood at the platform and picked apples from the tree, and the other worker stood on the ground to collect apples. There were four tests (four components to evaluate) in total, and for each test, 100 apples were collected (4 replications with 25 apples per replication). After each test, the experimental apples were kept in a room temperature for 12 h before they were peeled to obtain the bruising conditions, which was assessed following Table 2.1. In this study, if apples were bruised (class 2–7), they were recorded as bruised apples. For each component test, the apple bruising number was divided by the total amount (25 apples). The bruising ratio for each replication was obtained, and after the ratios of the four replications were averaged, it represented the bruising performance of the component.

Table 2.1 Classification of apple bruise damage [28]

Class	USDA fresh market standards	Bruise specification
1	"Extra Fancy"	No bruising
2	"Extra Fancy"	Bruise diameter ≤ 3.2 mm (1/8 in.)
3	"Extra Fancy"	Bruise diameter 3.2–6.4 mm (1/ 8 to 1/4 in.)
4	"Extra Fancy"	Bruise diameter 6.4 mm (1/4 in.) to 12.7 mm (1/2 in.) or area of several bruises < 127 mm^2
5	"Fancy"	Bruise diameter 12.7 to 19 mm (1/2 to 3/4 in.) or total area of multiple bruises < 283 mm^2
6	Downgraded	Bruises larger than the tolerances in "Fancy"
7	Downgraded	Cuts or punctures of any size

2.2.2.2 Second Year Field Test

Different from the first year experiment focusing on evaluating component for system improvement, the second year experiment targeted a systematic study on bruising and harvest efficiency. Improved distributor was integrated into the system for the second year test (Fig. 2.6b).

For the second year test, workers only harvested apples with a height about 210 cm (7 ft.) above the ground. For apples with the height > 210 cm, workers have to take advantage of ladders to reach and harvest. A plastic band was tied to the tree canopy as a reference for the height of 210 cm. A worker harvested apples for 15 consecutive trees in single, adjacent rows using ladder-and-bucket method, and the entire harvest process was video-recorded (HDR-CX550V high-definition handy camcorder, Sony Corp., Japan). The worker only picked apples on the side of trees facing him. For the harvest-assist approach, one picker stood on the platform to harvest 15 consecutive trees in the adjacent row. The harvest process was video recorded as well. The video was used to calculate the harvest efficiency (apples/s) and the time ratio pickers spending on harvesting.

2.3 Resutls and Discussion

2.3.1 First Year Field Test Results

The first year field test results of bruising caused by each component are shown in Fig. 2.7. The results show that the tube and manifold did not cause bruising, and the distributor and bin were determined as the sources for bruising. The high bruising caused by the bin could also be attributed to the distributor. Since the distributor was incapable of reducing the apple velocity to a low value, when the high velocity apples collided with bin surfaces, it led to bruising. The total bruising ratio was 49%

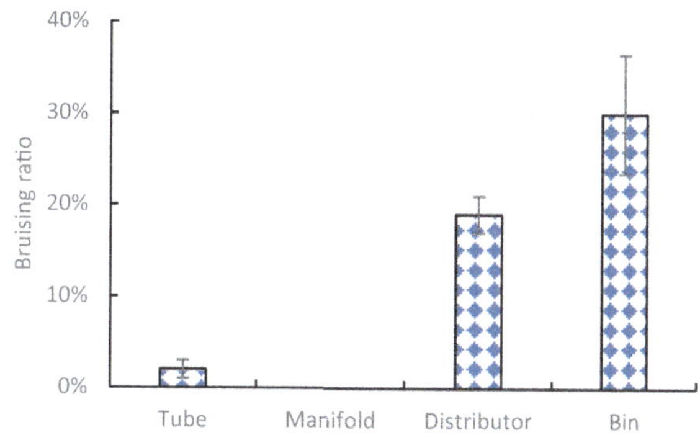

Fig. 2.7 First year bruising ratio of each component

(sum of ratios for the four components), which was far from meeting the fresh market requirements.

Those holes in the distributor were intended to distribute apples uniformly in the bin, which was based on the assumption that apples fall on the padding materials and then the centrifugal force would push apples from the center towards the edge. During the movement from the center to the edge, while apple go through the holes, they would fall through the holes into the bin. Since these holes having different distance to the disc center, it would assist the uniform distribution. However, during the careful observation of the field experiments, the conditions were different. In the field test, it was observed that some apples just fell through the hole into the bin after they exited from the manifold. Thus, the apples arrived at the bin with high velocity, and the deceleration function of the disc did not function. A worse scenario was when the apple falling through the hole, the hole edge hit the apple surface, which could lead to cut and puncture to the apple. It was noticed many cases that the cut and puncture conditions occurred during the experiment.

2.3.2 Second Year Field Test Results

Table 2.2 shows the time ratios spent in apple picking and harvest efficiency under two different harvest methods (e.g., ladder-bucket and harvest-assist unit) for the high level apples (>210 cm above ground). The time percentages spent in picking are closely related to the efficiency—more time spent on apple picking would result in

Table 2.2 Ladder-bucket and harvest-assist unit comparison for high level apple harvest (>210 cm above ground) in terms of time percentage spent in apple picking and harvest efficiency

Harvest method	Time percentage spent in apple picking (%)	Harvest efficiency (apples/s)
Ladder-bucket	76	0.20
Harvest-assist unit	98	0.39

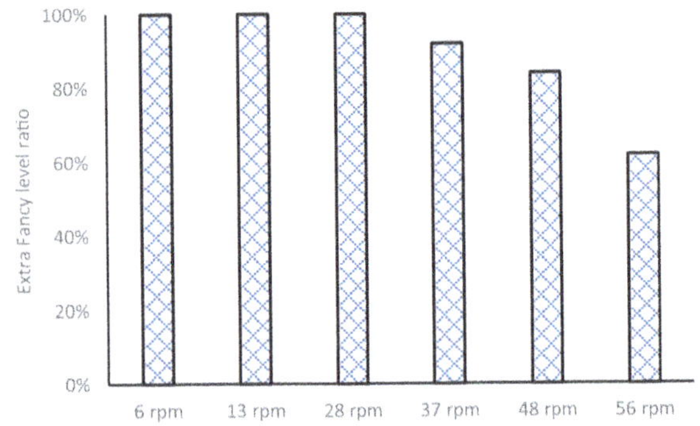

Fig. 2.8 Extra Fancy grade apple ratio under different distributor speeds; rpm represent revolution per minute

high efficiency. For the ladder-bucket approach, in addition to picking, workers have to conduct a lot of other activities that are irrelevant to apple picking, such as ladder moving, climbing/descending ladder, and transporting apples into the bin. However, when working on the harvest-assist unit, workers only need to drive the platform to the next picking location, which is convenient and consumes minimal time. In addition, when working on the unit, workers are free from apple transportation and dumping apples into the bin. The experimental results show that the time percentages for ladder-bucket and harvest-assist unit are 76% and 98%, respectively. The data reflects the real conditions that that workers using the harvest-assist unit would have more time in picking apples. The harvest efficiencies for ladder-bucket and harvest-assist unit are 0.20 and 0.39 apples/s, respectively. The harvest-assist unit is much higher efficiency over the ladder-bucket method.

Multiple distributor speeds were tested, including 6, 13, 28, 37, 48, and 56 rpm. The Extra Fancy level apple ratios under different distributor speeds are shown in Fig. 2.8. For the speeds of 6, 13, and 28 rpm the Extra Fancy levels are all 100%. With the distributor speed increases to 37, 48, and 56 rpm, the ratio of Extra Fancy level apples decreases, because the higher speed distributor would increase apple speed. Among the three speeds of 6, 13, and 28 rpm, though all generate minimal bruising to apples, it is recommended to use 28 rpm, as the slow speed have a chance to cause apple clogging.

2.4 Conclusion

A low-cost apple harvest-assist unit was designed, fabricated, and tested for the 2013 and 2014 harvest seasons. During the first year's test, it was observed that the multi-hole distributor did not perform satisfactorily due to the poor performance of the multi-hole distributor. Since the multi-hole distributor could not decelerate apples' speed satisfactorily and even cut/punctured apples, the second version of cone-shape bin filler was designed and incorporated into the low-cost apple harvest-assist unit. Field experiments showed that the time ratio spent in apple picking was 29% higher for the harvest-assist method over the ladder-bucket approach, because working on the unit, workers only need to pick apples and then drive the unit. However, when working with the ladder-bucket approach, workers have to conduct a lot of other activities that are irrelevant to apple picking, such as bending to dump apples into the bin.

Regarding the distributor, slow speed would cause apple clogging and high speed would increase the bruising ratio. After testing multiple speeds, it was concluded that 28 rpm performed more satisfactorily over other speeds (i.e., 6, 13, 37, 48, and 56 rpm). Thus, future study should use 28 rpm distributor spinning speed. The total cost of the machine is about $30,000: the ORSI eco-picking platform ~$25,000 and the harvest devices ~$5,000. Thus, it meets the original target of the machine total cost below $35,000.

There are still a lot of studies pending for this project. First, though the current version ORSI eco-pick platform is height adjustable, the approach is manual. It was observed in the field tests that under some extreme conditions, workers have to stretch their bodies to get access to apples. It is desirable to realize automatic height adjustment to benefit the apple picking process. Second, this study only tested the unit's performance on 'Fuji' apples, and more extensive tests on other apple cultivars should be conducted. Third, economic analysis of this machine should be conducted to provide information on apples growers who are interested in purchasing this machine.

Credit Authorship Contribution Statement Y. Shi: Writing—original draft. **Y. Wang**: Writing—review & editing. **Z. Zhang**: Conceptualization, Investigation, Methodology, Project administration, Writing—original draft, Writing—review & editing.

Disclaimer
Mention of commercial products or orchards in this paper is only for providing factual information and does not imply endorsement of them by authors over those not mentioned.

Declaration of Competing Interest
The authors declare that they have no known competing financialinterests or personal relationships that could have appeared to influence the work reported in this paper.

References

1. U.S. Apple Association (2021) Retrieved from https://usapple.org/
2. Zhang Z (2015) Design, test, and improvement of a low-cost apple harvest-assist unit. Ph.D. diss. University Park, PA: Pennsylvania State University, Department of Agricultural and Biological Engineering
3. Zhang Z, Heinemann PH, Liu J, Baugher TA, Schupp JR (2016a) The development of mechanical apple harvesting technology: a review. Trans ASABE 59(5):1165–1180. https://doi.org/10.13031/trans.59.11737
4. Zhang Z, Heinemann P, Liu J, Schupp J, & Baugher T (2017a) Brush mechanism for distributing apples in a low-cost apple harvest-assist unit. Appl Eng Agric 33(2):195–201. https://doi.org/10.13031/aea.1197

5. Calvin L, Martin P (2010) The US produce industry and labor: Facing the future in a global economy (No. 1477–2017–4011). Retrieved from https://ageconsearch.umn.edu/record/262245/
6. Zhang Z, Igathinathane C, Li J, Cen H, Lu Y, Flores P (2020) Technology progress in mechanical harvest of fresh market apples. Comput Electron Agric 175:105606. https://doi.org/10.1016/j.compag.2020.105606
7. Zhang Z, Pothula A, Lu R (2016c) Economic analysis of a self-propelled apple harvest and infield sorting machine for the apple industry. ASABE Paper No. 2456644. St. Joseph, MI: ASABE. https://doi.org/10.13031/aim.20162456644
8. Zhang Z, Pothula A, Lu R (2017b) Economic evaluation of apple harvest and infield sorting technology. Trans ASABE 60(5):1537–1550. https://doi.org/10.13031/trans.12226
9. Gongal A, Amatya S, Karkee M, Zhang Q, Lewis K (2015) Sensors and systems for fruit detection and localization: A review. Comput Electron Agric 116:8–19. https://doi.org/10.1016/j.compag.2015.05.021
10. Zhang Z, Heinemann P (2017c) Economic analysis of a low-cost apple harvest-assist unit. HortTechnology 27(2):240–247. https://doi.org/10.21273/horttech03548-16
11. Berlage AG (1973) Apple harvesting trials with oscillating air jets. Trans ASAE. 16(3):460–461. https://doi.org/10.13031/2013.37542
12. Le Flufy MJ (1982) The design of a prototype apple harvester. J Agric Eng Res 27(1):51–60. https://doi.org/10.1016/0021-8634(82)90057-9
13. Le Flufy MJ (1982) Harvest trials with a prototype apple harvester. J Agric Eng Res 27(5):415–420. https://doi.org/10.1016/0021-8634(82)90079-8
14. Le Flufy MJ (1983) Apple harvesting by a combing technique. Trans ASAE 26(3):661–664. https://doi.org/10.13031/2013.33997
15. Millier WF, Rehkugler GE, Pellerin RA, Throop JA, Bradley RB (1973) Tree fruit harvester with insertable multilevel catching system. Trans ASAE 16(5):844–850. https://doi.org/10.13031/2013.37641
16. Peterson DL, Bennedsen BS, Anger WC, Wolford SD (1999) A systems approach to robotic bulk harvesting of apples. Trans ASAE 42(4):871–876. https://doi.org/10.13031/2013.13266
17. Thomas RL (1964) An engineering investigation in the use of a pulsating air stream to mechanically detach apples from the tree. MS Thesis. Rutgers University, New Brunswick, NJ
18. Davidson JR, Mo C (2015) Mechanical design and initial performance testing of an apple-picking end-effector. ASME Int Mech Eng Congr Exposition. Am Soc Mech Eng. 1–9. https://doi.org/10.1115/imece2015-50482
19. Schupp J, Baugher T, Winzeler E, Schupp M, Messner W (2011) Preliminary results with a vacuum assisted harvest system for apples. Fruit Notes 76(4):1–5
20. Zhang Z, Zhang Z, Wang X, Liu H, Wang Y, Wang W (2019) Multi-purpose apple harvest platform economic evaluation modeling and software development. Int J Agric Biol Eng 12(1):74–83. https://doi.org/10.25165/j.ijabe.20191201.4360
21. Fathallah FA (2010) Musculoskeletal disorders in labor-intensive agriculture. Appl Ergon 41(6):738–743. https://doi.org/10.1016/j.apergo.2010.03.003
22. Lu R, Zhang Z, Pothula AK (2017) Innovative technology for apple harvest and in-field sorting. Fruit Quart 25(2):11–14
23. Robinson T, Sazo MM (2013) Recent advances of mechanization for the tall spindle orchard system in New York State – Part 2. NY Fruit Q 21(3):3–7
24. Robinson T, Hoying S, Sazo MM, DeMarree A, Dominguez L (2013) A vision for apple orchard systems of the future. New York Fruit Q 21(3):11–16
25. Luo R, Lewis KM, Zhang Q, Wang S (2012) Assessment of bruise damage by vacuum apple harvester using an impact recording device. ASABE Paper No. 121338094, St. Joseph, MI: ASABE. https://doi.org/10.13031/2013.41870
26. Zhang Z, Heinemann PH, Liu J, Schupp JR, Baugher T A (2016b) Design and field test of a low-cost apple harvest-assist unit. Trans ASABE 59(5):1149–1156. https://doi.org/10.13031/trans.59.11708

27. Zhang Z, Heinemann P, Liu J, Schupp J, Baugher T (2014) Design, fabrication, and testing of a low-cost apple harvest-assist device. ASABE Paper No. 141839738, St. Joseph, MI: ASABE. https://doi.org/10.13031/aim.20141839738
28. Peterson DL, Tabb AL, Baugher T, Lewis K, Glenn DM (2010) Dry bin filler for apples. Appl Eng Agric 26(4):541–549. https://doi.org/10.13031/2013.32057

Chapter 3
Economic Evaluation of a Low-Cost Fresh Market Apple Harvest-Assist Unit

Zhaohua Zhang, C. Yang, Y. Wang, and Zhao Zhang

Abstract To replace the inefficient and occupational disease-prone conventional ladder-bucket apple harvest approach, a low-cost apple harvest-assist unit was developed. This study conducted economic analysis on this unit to provide baseline information for its commercialization. The unit brings benefits to apple growers in four aspects: improved harvest efficiency, decreased occupational injuries, eliminated ladder fees, and increased working efficiency on pruning, training, and thinning (PTT). The economic analysis was conducted at the yields between 25 to 45 Mg/ha, which covered the U.S. apple orchard yield ranges. Savings on PTT and improved harvest efficiency account for 61% and 33% of the total savings, which indicates the importance of using the unit for multiple purposes. For any apple orchards with an area smaller than 4.2 ha, it is not recommended to purchase the unit, because the adoption would result in negative net benefits to apple growers. For larger apple orchards, apple growers are recommended that apple growers to purchase multiple units to meet the needs. Taking a 23 ha apple orchard with a yield of 25 Mg/ha for

Z. Zhang
College of Economics and Management, Shandong Agricultural University, Tai'an 271018, China
e-mail: zzhang@sdau.edu.cn

C. Yang
Department of Bioproducts and Biosystems Engineering, University of Minnesota, Saint Paul, MN 55108, USA
e-mail: ceyang@umn.edu

Y. Wang
Agricultural Information Institute of Chinese Academy of Agricultural Sciences, Beijing 100125, China
e-mail: wangyuanjie@caas.cn

Z. Zhang (✉)
Key Lab of Modern Precision Agriculture System Integration Research, Ministry of Education of China, China Agricultural University, Beijing 100083, China
e-mail: zhang.zhao.1@ndsu.edu

Key Lab of Agriculture Information Acquisition Technology, Ministry of Agriculture and Rural Affairs of China, China Agricultural University, Beijing 100083, China

Department of Agricultural and Biosystems Engineering, North Dakota, State University, Fargo, ND 58102, USA

Z. Zhang et al. (eds.), *Mechanical Harvest of Fresh Market Apples*, Smart Agriculture 1,
https://doi.org/10.1007/978-981-16-5316-2_3

example, purchase of two units to meet the practical requirements would help apple growers save $8000 per year. This study has provided baseline information to assist apple growers in decision making on investment of the harvest-assist unit based on the orchard size and yield.

Keywords Harvest-assist · Platform · Low-cost · Economic analysis · Fresh market apples · Decision making

3.1 Introduction

As one of the most popular fruits worldwide, apples rank the No. 1 consumed fruit in the U.S. [1]. Across the U.S., apples are grown in more than 5000 orchards in 32 states, and the farm-gate revenue of wholesale value is about $\$5 \times 10^9$ [2]. Though apples are consumed in a number of manners, such as fresh eating, juice, canned, dried, and sauce, more than two thirds of apples are consumed in fresh eating. Due to the high quality requirement (minimal bruising, large size, and good appearance), susceptibility to bruising, and lack of appropriate technologies, apples are still manually harvested [3–6]. During the harvest window, seasonal workers use ladder-bucket approach to harvest apples. For the low-level apples, workers pick on the ground,while for the high-level apples, workers need take advantage of ladders to get access to them [7, 8]. Throughout the entire harvest process, workers have to wear a bucket to temporarily hold apples, and when the bucket is full, workers walk to a large bin to release apples, after which workers return to the tree to continue with apples picking [6, 9, 10]. There are many disadvantages of the ladder-bucket method. First, the overall harvest process is inefficient, as workers are involved with a number of activities that are indirectly related to apple picking, such as moving ladders, and transporting apples into the bin [11, 12]. Second, the harvest work is strength demanding, because workers need to wear the bucket to hold apples. When the bucket is full, it weighs about 20 kg, and even worse, workers have to wear such a heavy bucket to conduct other activities, such as climbing and descenting the ladder. Due to the high requirement for physical strength, some seasonal workers are incapable of conducting the harvest work, which worsens the labor shortage issue [10, 13]. Third, the ladder-bucket approach is prone to causing occupational injuries. While reaching apples, regardless of standing on the ground or ladder, workers always use awkward postures to get access to apples to expedite the harvest process, as they are paid on a piece rate. Using awkward postures would easily result in occupational diseases (e.g., strain and sprain) [14]. In addition, while releasing apples into the bin, workers need to bend slowly to avoid bruising, during which process it has a chance to cause occupational injuries on the back [15]. Due to existing disadvantages associated with the ladder-bucket harvest method, researchers initiated different projects to develop and test new fresh market apple harvest methods [6, 8, 16, 17].

Researchers first explored the potential of using bulk harvesting approach for fresh market apple harvesting. The bulk harvesting uses trunk shaking, fruit pressing, or air jets to detach multiple apples from tress simultaneously and automatically [18–21]. The detached apples are collected by catching mechanisms and then transported to a bin [22, 23]. Though multiple prototypes were developed, tested, and optimized, none came to the commercial stage due to excessive bruises and high technology cost [6, 24]. Since the bulk harvesting could not meet the practical application requirement, researchers shifted to the development and integration of apple harvest robots [25, 26]. Instead of detaching multiple apples at a time, the robot detaches one apple at a time [27]. Developing robots needs knowledge of different disciplines, such as image vision, mechanical design, control, circuit design, and embedded systems. Several apple harvest robots were developed and tested, but none came to the commercial stage for a number of reasons, such as high cost, low technology reliability, and high apple bruising [8, 28, 29]. Since the application of ladders has been verified as to be the major cause of harvest inefficiency and occupational injuries, researchers started to find an innovative way to replace the use of ladders, which resulted in the idea of harvest-assist platform concept [30]. By standing on the platform to pick apples, there is no need to use ladders, and this further gets workers rid of ladder-associated activities. So far, several harvest platforms are commercially available on the market.

For all commercial harvest platforms, they are all large in size, and have a high price. For large apple orchards, these platforms are acceptable, as the technology adoption would bring benefits. However, for small orchards, apple growers are hesitant in investing the existing platforms for the high price and applicability in small orchards [9, 10]. To meet the technology requirement for small apple orchards, researchers at The Pennsylvania State University developed a low-cost apple harvest-assist unit [11]. The unit was integrated, and field tests had demonstrated its potential for commercial application, with satisfactory performance on minimal apple bruising and significant harvest efficiency increase. Though the harvest-assist unit is technologically sound, its economic performance would finally decide if apple growers are willing to adopt the technology. In general, if the machine costs are higher than the savings, apple growers would have less interest in purchasing the machine,while if the machine costs are lower than the savings, there is a huge potential that apples growers would purchase the machine.

A critical goal of the low-cost harvest-assist unit project is to commercialize the research outcomes. There are a lot of agricultural machinery that have been developed, but there are only a few economic evaluations conducted. Objective economic evaluation would provide baseline information on the commercial potential of the harvest-assist unit. To evaluate the economics of the harvest-assist unit, specific objectives of this study are to: (1) calculate the cost of adopting the harvest-assist unit; (2) estimate cost savings by using the unit; and (3) compare the difference between cost and savings. Furthermore, this study would provide some related information that other researchers can refer to when they assess economic performance of other newly developed machines.

3.2 Materials and Methods

3.2.1 Brief Introduction to the Low-Cost Apple Harvest-Assist Unit

The low-cost fresh market apple harvest-assist unit consists of two major parts: mobile platform and harvest device. An ORSI Eco-pick self-propelled mobile platform (ORSI Group, Bologna, Italy) was purchased for this study, which was used as the base to mount the harvest device. The mobile platform is able to accommodate 2 workers, and powered by a set of 12 V DC batteries, which could last about 5 ~ 6 h. Since there are two sets of batteries, and when one set is in use, the other set can be charged. The alternation of two sets of batteries would minimize the downtime. The mobile platform is manually adjustable both horizontally and vertically. When the platform is fully extended, its width is about 2.4 m, and when fully retracted, the width is about 1.2 m. The adjustable width makes it capable of accommodating different apple tree row distance. The adjustable height makes the platform accommodate different height of trees—for orchards with tall trees, the platform can be adjusted a little higher, so workers can get access to the high level apples. The platform height (where the workers stand on in Fig. 3.1) is adjustable from 90 to 160 cm to the ground.

The harvest device consists of receiver, tube, manifold, and distributor. After apples are picked from trees, workers put them into the receiver. Apples are first transported by the tube to the manifold, and then they arrive at the distributor. The

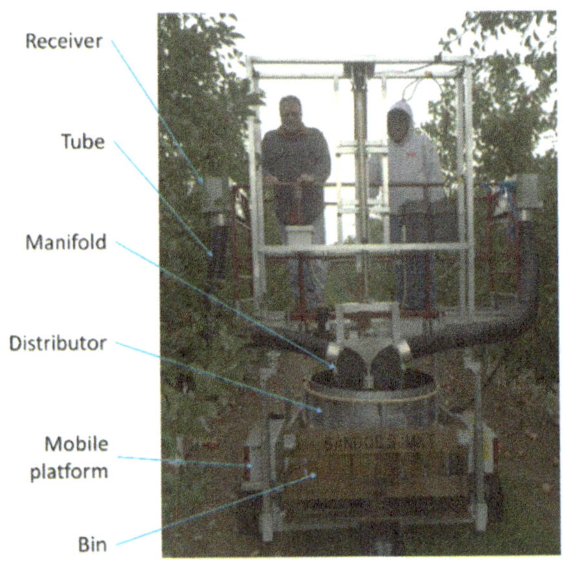

Fig. 3.1 View of the low-cost fresh market apple harvest-assist platform in field test

distributor collects apples and then distributes them into the bin evenly with minimal bruising or without bruising. Details of the machine design and parameters can be referred to Zhang et al. [9, 10].

3.2.2 *Basic Parameters Used for Economic Analysis*

Considering the apple harvest window is a little over one month, the unit was assumed to operate 45 days with 8 h/day per harvest season. In addition, to improve the economic returns, the unit was assumed to operate a total of 100 days with 8 h/day on pruning, training, and thinning (PTT) [31]. Thus, the unit would be used a total of 1160 h (360 h for harvesting and 800 h for PTT) yearly. It is estimated that the life cycle of the unit to be 8 years, based on which the unit would operate a total of 9280 h (1160 h/year × 8 year) in the life cycle. The unit consists of two parts: the ORSI platform and the harvest device. The ORSI platform was purchased at the price of $15,000 and the total cost (including labor and materials) for the harvest device was about $10,000. After adding a $5000 profit for dealer, the unit price was estimated to be $30,000. Both the ORSI platform and the harvest device are powered by 12 V DC batteries. The electronic components in the harvest device would consume the energy, including motors to spin the distributor, sensors to detect the apple height in the bin, and micro-controller to adjust the distributor speed. The $30,000 price include backup batteries to avoid the downtime, and these batteries were assumed to function during the life cycle of the unit.

3.2.3 *Annual Costs of the Unit*

Annual costs of the harvest-assist unit comprise two sections: annual ownership and operational costs [32–34]. The annual ownership costs are also named as fixed costs, because they occur even without use of the machine. The annual ownership cost consists of depreciation, interest, taxes, housing, and insurance. On the opposite side, the annual operational costs are directly related to the service time of the machine.

3.2.3.1 Annual Ownership Costs

The annual ownership costs start to occur after the machine is purchased, and end when the machine is sold or no longer operable [35]. Depreciation reflects the machine value reduction, among which age and accumulated hours are the two major contributing parameters. Salvage value, representing the sale price of the machine after a certain amount of time, is another crucial factor affecting depreciation. The salvage value can be estimated by multiplying the machine purchase price by a salvage value factor. Based on the assumption that the unit has an 8-year life cycle, the salvage value factor was reasonably assumed to be 0.1 in this study [36, 37].

Following depreciation, interest is another important factor determining the ownership cost. If the unit purchasing money comes from bank loan, the interest rate is determined by the bank; if the money comes from apple growers' own capital, the interest would be determined by the opportunity cost. Considering a majority of the apple growers use bank loan to purchase orchard equipment, this study reasonably assumes all the investment comes from the bank, and the interest rate was assumed to be 8% [38].

In addition to depreciation and interest, taxes, insurance and housing (TIH) are three other contributing factors to ownership costs. TIH are usually estimated as the percentage of the machine purchasing price [12]. According to ASABE standards [33], TIH was assumed as 2% of the machine purchasing price: taxes, insurance, and housing as 1.00%, 0.25%, and 0.75%, respectively.

The ownership costs can be calculated based on the Eqs. 3.1 and 3.2. C_A is the annual ownership costs, and P_M and C_0 are the new machine purchase price ($30,000) and the ownership cost coefficient. S_V stands for the machine salvage value factor (0.1), and L represents the machine life cycle (8 years). The annual interest rate is denoted by i (8%), and K_2 is the TIH (2%).

$$C_A = P_M \times C_0. \tag{3.1}$$

$$C_0 = \frac{1 - S_V}{L} + \frac{1 + S_V}{2} \times i + K_2 \tag{3.2}$$

3.2.3.2 Operational Costs

Annual operational costs include repairs, maintenance, and fuel. The repairs and maintenance costs can be calculated using Eq. 3.3, in which C_{rm} represents the annual repair and maintenance costs. RF_1 and RF_2 represent the repair and maintenance factors one and two, respectively. h is the accumulated annual use of machine (h). Based on the ASABE standards [32], RF_1andRF_2 values were selected as 0.03 and 2, respectively.

$$C_{rm} = P_M \times RF_1 \times \left[\frac{h}{1000}\right]^{RF_2} \tag{3.3}$$

As the harvest-assist unit is fully powered by electricity, fuel cost means electricity cost. The mobile platform consumes 1600 W power. Other components consuming electricity include a linear actuator (100 W) to lift the manifold-distributor and a gear-motor (130 W) to spin the distributor. The other two electronic parts consuming electricity are the sensor to detect the apple height in the bin and the controller for distributor spinning speed adjustment. Since these two electronic parts consume very little energy, it was assumed to be a total of 1 W in this study. All energy was provided by rechargeable 12 V DC batteries, and as during the charging process, part of the energy would be wasted, thus, a coefficient of 1.05 was used. The electricity cost was estimated to be \$0.18/kWh [39].

3.2.3.3 Machine Capacity and Machine Coverage Area

Machine capacity (MC) is defined as kilograms of apples the unit can harvest in one harvest season, and the coverage area represents orchard area can be covered by one unit in one harvest season. The MC could be obtained by Eq. (3.4), where H_{tp}, THT, and W_a represents harvest-assist unit throughput (0.78 apples/s), total harvest time (1.3×10^6 s), and weight of an apple (0.15 kg/apple), respectively [7]. The machine coverage area can be calculated using Eq. (3.5), where W_a and 0.5 denotes the orchard yield (ton per hectare) and 50% of apples in a tree are harvested by the unit (high level apples), respectively [7]. Based on these parameters and Eqs. (3.4 and 3.5), the annual MC is calculated as 1.5×10^5 kg.

$$MC = H_{tp} \times THT \times \frac{W_a}{1000} \tag{3.4}$$

$$MCA = \frac{MC}{0.5 \times OY} \tag{3.5}$$

3.2.4 *Annual Cost Savings by the Unit*

Adoption of the low-cost apple harvest-assist unit would generate cost savings to apple growers in terms of improving apple harvest efficiency, decreasing occupation injuries, increasing working efficiency on PTT, and avoiding ladder fees (purchase and maintenance) [12].

3.2.4.1 Increased Apple Harvest Efficiency

After the apple harvest-assist unit was developed, it was extensively tested in The Pennsylvania State Fruit Research and Extension Center. The experimental results demonstrated that for the high level apples (traditionally harvested by using ladders), the adoption of the harvest-assist unit could improve the harvest efficiency by 95%. Standing on the platform, each worker picks apples at the throughput of 0.39 apple/s, and since two workers pick simultaneously, the H_{tp} is 0.78 apples/s.

3.2.4.2 Decreased Occupational Injuries

The harvest efficiency increase (EI) can be calculated using (Eq. 3.6). In Eq. (3.6), EI, RE, ER, OII, and SI represents efficiency increase, reduced efficiency by occupational injuries, reduced occupational injury percentage, occupational injury incidence, and sprain/strain incidence of back, neck, and shoulder.

$$EI = \left(\frac{1}{1 - RE} - 1\right) \times ER \times OII \times SI \tag{3.6}$$

Fulmer et al. [40], Earle-Richardson et al. [41, 3, 4], Brower et al. [42] and conducted different studies and have demonstrated that the reduced efficiency by occupational injuries roughly ranging from 5 to 20%. In this study, RE was reasonably estimated to be 15%. The ER value was chosen as 34% based on [11]. The OII and SI was selected as 5% and 37%, respectively [42, 40]. According to all the provided parameters, the EI was calculated as 0.7%. However, there were a lot of other serious conditions (e.g., death and fractures caused by ladder falls) that could not be quantified, so they could not be incorporated into Eq. 3.6. Thus, the EI was reasonably adjusted as 2% [12].

3.2.4.3 Increased Working Efficiency on PTT

Growers can take advantage of the harvest-assist unit to conduct PTT to generate more cost savings. Due to time limitation and funding constrains, the performance of the developed unit on improving working efficiencies on PTT has not been tested yet. However, other researchers [43–45] conducted PTT efficiency improvement on platforms similar to the developed harvest-assist unit. Based on reviewing their experimental results, the PTT efficiency was estimated to increase 50% via applying the harvest-assist unit in this study.

3.2.4.4 Eliminating Expenditures on Ladder Purchase and Maintenance

By fully replacing ladders, apple growers no longer need to invest ladders for the harvest and PTT use. Thus, the purchase cost of ladder and associated maintenance cost are saved. Based on the study results by Gallardo and Pedroso-Galinato 46, it was estimated that an apple orchard requires \$39.6/ha ladder fee for a yield of 45 Mg/ha, which would be fully saved by adopting the harvest-assist unit.

3.2.5 Orchard Area Threshold

While investing in the harvest-assist unit, apple growers would mainly make their decisions based on the analysis results. If the annual costs are less than the annual savings (positive net benefits), apple growers would be willing to purchase the unit (the unit would have market potential). However, if the annual costs are more than the annual savings (negative net benefit), apple growers would have less interest in purchasing the unit. The orchard area is a critical factor in determining the net benefit by using the harvest-assist unit: for a very small orchard, the net benefit would be very low or even negative, as the unit is not used much. In addition, the orchard yield would affect the determination of area threshold—the area that the machine cost equals to cost savings (net benefit equals to 0). There are four major apple production states in the U.S.—Washington, New York, Michigan, and Pennsylvania. For these four states, apple yields range from 2.5×10^4 to 4.5×10^4 kg/ha [13]. Thus, 2.5×10^4, 3.0×10^4, 3.5×10^4, 4.0×10^4, and 4.5×10^4 orchard yields would be used as case studies.

3.3 Resutls and Discussion

3.3.1 Cost Savings from Increased Harvest Efficiency

Gallardo and Galinato [46], Seavert et al. [31] and Zhang and Heinemann [12] reported that when the yield was 45 Mg/ha, the harvest time was 120.7 h/ha, based on which the harvest time under different yields was calculated. Since the harvest efficiency improvement was 29%, the harvest time saved by using harvest-assist unit was calculated. After confirming the harvest time saved by using harvest-assist unit, the cost savings were calculated according to the assumption that harvest labor was \$20/h. The machine covered area was calculated using Eq. (3.5). All the calculated results are shown in Table 3.1. With the yield increases from 25 to 45 Mg/ha, the cost savings increases from 301 to 525 \$/ha. However, the machine covered area decreases as yield increases, because the machine capacity is a constant value (300 Mg/ha).

3.3.2 Cost Savings from Decreased Occupational Injuries

Based on the reported data that when the yield was 45 Mg/ha, the harvest time was 120.7 h, and the finally calculated EI value of 2%, the cost savings by decreasing occupational injuries under different yields were shown in Table 3.2. For yield ranging from 25 to 45 Mg/ha, the labor cost savings ranging from 26 to 47 \$/ha. The higher the yield, the more the saving.

Table 3.1 Harvest cost savings by replacing ladder-bucket method using an apple harvest-assist unit

Yield (Mg/ha)	Harvest time per ha using ladder-bucket approach (h)	Saved harvest time per ha using harvest-assist unit (h)[a]	Labor cost savings per ha ($)[b]	Machine covered area (ha)[c]
25	67.1	15.1	301.7	12.0
30	80.5	18.1	361.9	10.0
35	93.9	21.1	422.2	8.6
40	107.3	24.1	482.4	7.5
45	120.7	27.1	524.7	6.7

[a] Harvest efficiency improvement of 29%; [b] Harvest labor cost of $20/h; [c] Machine capacity of 300 Mg/ha per harvest season

Table 3.2 Cost savings from decreasing occupational injuries under different orchard yields

Yield (Mg/ha)	Harvest time per ha using ladder-bucket approach (h)	Saved harvest time per ha using harvest-assist unit (h)[a]	Labor cost savings per ha ($)[b]	Machine covered area (ha)[c]
25	67.1	1.3	26.3	12.0
30	80.5	1.6	31.6	10.0
35	93.9	1.8	36.8	8.6
40	107.3	2.1	42.1	7.5
45	120.7	2.4	47.3	6.7

[a] Harvest efficiency improvement of 2%;[b] Harvest labor cost of $20/h; [c] Machine capacity of 300 Mg/ha per harvest season

3.3.3 Cost Savings from Improved Working Efficiencies in PTT

Previous studies showed that when an orchard yield is 45 Mg/ha, it required 261 h to complete the PTT activities [12, 31, 46]. Considering the adoption of the low-cost harvest-assist unit would increase harvest efficiency by 50% and the PTT labor is $11.5/h, the cost savings on PTT were calculated (Table 3.3). While the yield is 25 Mg/ha, the labor savings are $555/ha, and when the yield goes up to 45 Mg/ha, the savings are $999/ha.

Table 3.3 PTT cost savings for increase labor efficiency by using harvest-assist unit under different orchard yields

Yield (Mg/ha)	PTT time per ha using ladder-bucket approach (h)	Saved time on PTT per ha using harvest-assist unit (h)[a]	Labor cost savings per ha ($)[b]	Machine covered area (ha)[c]
25	144.8	48.3	555.2	12.0
30	173.8	57.9	666.2	10.0
35	202.8	67.6	777.3	8.6
40	231.7	77.2	888.3	7.5
45	260.7	86.9	999.4	6.7

[a] Harvest efficiency improvement of 50%; [b] Labor cost of $11.5/h; [c] Machine capacity of 300 Mg/ha per harvest season; PTT: pruning, training, and thinning

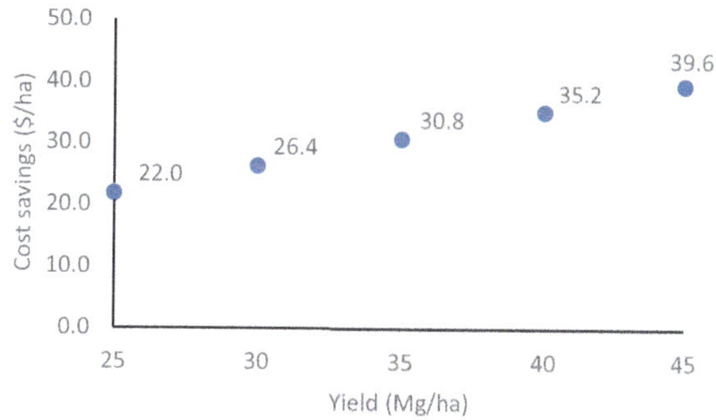

Fig. 3.2 Ladder cost savings by adoption of harvest-assist unit under different orchard yields

3.3.4 *Cost Savings from Avoiding Ladder Use*

By adopting the harvest-assist unit, ladder use is removed, and the savings were caused by avoiding the ladder purchase cost and the associated maintenance fee. Since an orchard with a yield of 45 Mg/ha would require a cost of $39.6/ha fee for ladder, the savings on ladders under different yield are shown in Fig. 3.2. With the yield increases from 25 to 45 Mg/ha, the cost savings range from $ 22.0/ha to $39.6/ha.

3.3.5 *Net Benefits Based on Orchard Area and Yields for a Harvest-Assist Unit*

The benefits and orchard area threshold for one harvest-assist unit under different yields are shown in Fig. 3.3. With the orchard yield increases from 25 to 45 Mg/ha, the orchard area threshold decreases from 7.6 to 4.2 ha. When the orchard yield

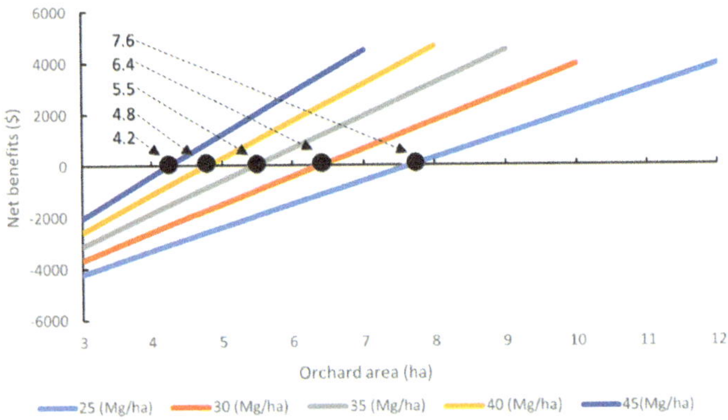

Fig. 3.3 Benefits and orchard area threshold for one harvest-assist unit under different yields

is high, the unit is used more for a certain area of an orchard (e.g., per ha). With more hours used of the unit, the orchard area threshold declines. For any orchards with area smaller than 4.2 ha, it is not recommended to use the harvest-assist unit, because it would generate negative benefits to apple growers. On the opposite side, it is encouraged to invest on the unit for an orchard with an area larger than 7.6 ha, because the investment would bring positive benefits to apple growers.

3.3.6 Unit Quantity Needed and Benefits for Different Orchard Areas and Yields

Figure 3.3 shows the benefits of one harvest-assist unit, and it can be observed that one unit can only cover a relatively small area (7~12 ha). It lacks the information if an apple grower has an orchard larger than 12 ha. Economic benefits of multiple harvest-assist units under different orchard areas and yields are shown in Fig. 3.4. Generally, with more harvest-assist units adopted, the more economic benefits can be obtained. For an orchard with a yield of 25 Mg/ha and orchard area is 23 ha, it is recommended to purchase two harvest-assist units, and annual economic benefits are about $8,000.

A case study is further conducted: orchard yield of 40 Mg/ha and orchard size of 6.5 ha. The major benefits come from using the unit on PTT (61%), and this stress the importance to use a unit for multiple purposes. Following the PTT, the cost savings from improving harvest efficiency accounts for 1/3 of the total benefits. The benefits from decreasing occupational injuries and eliminating ladder use account for 3% and 2% of the total benefits, respectively (Fig. 3.5).

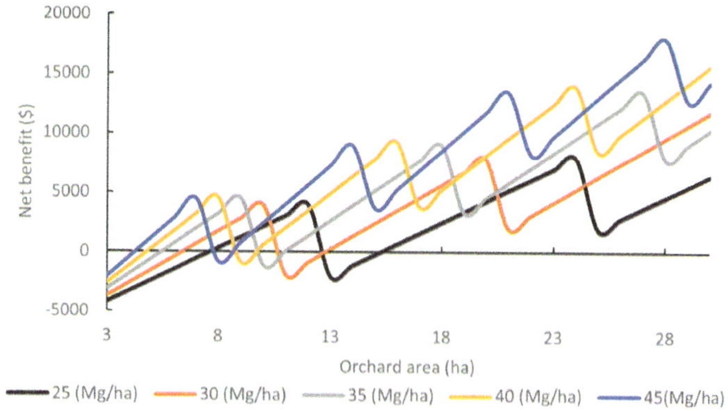

Fig. 3.4 Benefits of multiple harvest-assist units with different areas and yields

Fig. 3.5 Benefit ratios of different sources for a case study: orchard size 6.5 ha, yield 40 Mg/ha. PTT: pruning, training, and thinning

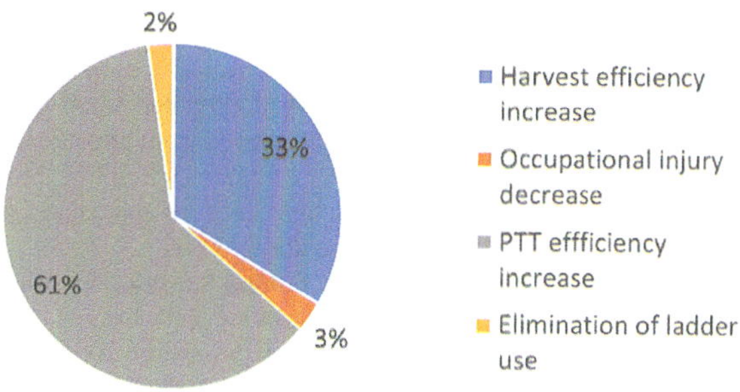

3.4 Conclusion

After developing and validating the low-cost apple harvest-assist unit technology, this study conducts economic benefit analysis of the unit, which can be used as a reference by apple growers to decide the investment. The harvest-assist unit would bring benefits to apple growers in terms of improving harvest efficiency, reducing occupational injury, increasing pruning, training, and thinning (PTT) efficiency, and eliminating ladder use. Among the four aspects, the savings by increasing PPT account for 61% of the total savings, followed by increasing harvest efficiency (33%). Thus, it is critical to have the unit perform multi-functions to bring more benefits to apple growers. The increased benefits would help apple growers adopt the new technology. It therefore should stress the multiple functions of the unit while introducing the unit to apple growers. Since adoption of the unit would bring negative benefits to any orchards with an area smaller than 4.2 ha, apple orchards with area smaller than 4.2 ha should not be targeted for this technology. For very lager orchards, growers are recommended to purchase multiple units to satisfy their needs. The economic analysis results would help better marketing the low-cost apple harvest-assist unit, and provide necessary information to assist apple growers in decision making on purchasing the machine.

Credit Authorship Contribution Statement **Z. Zhang**: Writing—original draft. **C. Igathinathane**: Writing—review & editing. **C. Yang**: Writing—review & editing. **Y. Wang**: Writing—review & editing. **Z. Zhang**: Conceptualization, Investigation, Methodology, Project administration, Writing—original draft, Writing—review & editing.

Disclaimer
Mention of commercial products or orchards in this paper is only for providing factual information and does not imply endorsement of them by authors over those not mentioned.

Declaration of Competing Interest
The authors declare that they have no known competing financialinterests or personal relationships that could have appeared to influence the work reported in this paper.

References

1. USDA (2018) Apples and oranges are America's top fruit choices. Retrieved from https://www.ers.usda.gov/data-products/chart-gallery/gallery/chart-detail/?chartId=58322
2. U.S. Apple Association (2021) Retrieved from https://usapple.org/
3. Earle-Richardson G, Jenkins PL, Strogatz D, Bell EM, May JJ (2006) Development and initial assessment of objective fatigue measures for apple harvest work. Appl Ergon 37(6):719–727. https://doi.org/10.1016/j.apergo.2005.12.002
4. Earle-Richardson G, Jenkins P, Fulmer S, Mason C, Burdick P, May J (2005) An ergonomic intervention to reduce back strain among apple harvest workers in New York State. Appl Ergon 36(3):327–334. https://doi.org/10.1016/j.apergo.2004.12.003
5. Freivalds A, Park S, Lee C, Earle-Richardson G, Mason C, May JJ (2006) Effect of belt/bucket interface in apple harvesting. Int J Ind Ergon 36(11):1005–1010. https://doi.org/10.1016/j.ergon.2006.08.005
6. Zhang Z, Heinemann PH, Liu J, Baugher TA, Schupp JR (2016a) The development of mechanical apple harvesting technology: a review. Trans ASABE 59(5):1165–1180. https://doi.org/10.13031/trans.59.11737
7. Zhang Z, Heinemann P, Liu J, Schupp J, Baugher T (2014) Design, fabrication, and testing of a low-cost apple harvest-assist device. ASABE Paper No. 141839738, ASABE, St. Joseph, MI. https://doi.org/10.13031/aim.20141839738
8. Zhang Z, Igathinathane C, Li J, Cen H, Lu Y, Flores P (2020) Technology progress in mechanical harvest of fresh market apples. Comput Electron Agric 175:105606. https://doi.org/10.1016/j.compag.2020.105606
9. Zhang Z, Heinemann PH, Liu J, Schupp JR, Baugher TA (2016b) Design and field test of a low-cost apple harvest-assist unit. Transactions ASABE 59(5):1149–1156. https://doi.org/10.13031/trans.59.11708
10. Zhang Z, Heinemann P, Liu J, Schupp J, Baugher T (2017a) Brush mechanism for distributing apples in a low-cost apple harvest-assist unit. Appl Eng Agric 33(2):195–201. https://doi.org/10.13031/aea.1197
11. Zhang Z (2015) Design, test, and improvement of a low-cost apple harvest-assist unit. Ph.D. diss. University Park, PA: Pennsylvania State University, Department of Agricultural and Biological Engineering
12. Zhang Z, Heinemann P (2017c) Economic analysis of a low-cost apple harvest-assist unit. HortTechnology 27(2):240–247. https://doi.org/10.21273/horttech03548-16
13. Zhang Z, Pothula A, Lu R (2017b) Economic evaluation of apple harvest and infield sorting technology. Trans ASABE 60(5):1537–1550. https://doi.org/10.13031/trans.12226

14. Zhang Z, Zhang Z, Wang W, Liu H, Sun Z (2019) The role of a new harvest platform in alleviation of apple workers' occupational injuries during harvest. J Agric Saf Health 25(1):11–24. https://doi.org/10.13031/jash.13103
15. Fathallah FA (2010) Musculoskeletal disorders in labor-intensive agriculture. Appl Ergon 41(6):738–743. https://doi.org/10.1016/j.apergo.2010.03.003
16. Gongal A, Amatya S, Karkee M, Zhang Q, Lewis K (2015) Sensors and systems for fruit detection and localization: a review. Comput Electron Agric 116:8–19. https://doi.org/10.1016/j.compag.2015.05.021
17. Zhao Y, Gong L, Huang Y, Liu C (2016) A review of key techniques of vision-based control for harvesting robot. Comput Electron Agric 127:311–323. https://doi.org/10.1016/j.compag.2016.06.022
18. Allshouse GW, Morrow CT (1972) Over-the-row harvester for dwarf fruit trees. Trans ASAE 15(6):1038–1043. https://doi.org/10.13031/2013.38067
19. Le Flufy MJ (1982a) The design of a prototype apple harvester. J Agric Eng Res 27(1):51–60. https://doi.org/10.1016/0021-8634(82)90057-9
20. Le Flufy MJ (1982b) Harvest trials with a prototype apple harvester. J Agric Eng Res 27(5):415–420. https://doi.org/10.1016/0021-8634(82)90079-8
21. Le Flufy MJ (1983) Apple harvesting by a combing technique. Trans ASAE 26(3):661–664. https://doi.org/10.13031/2013.33997
22. Millier WF, Rehkugler GE, Pellerin RA, Throop J A, Bradley RB (1973) Tree fruit harvester with insertable multilevel catching system. Trans ASAE 16(5):844–850. https://doi.org/10.13031/2013.37641
23. Peterson DL, Wolford SD (2003) Fresh–market quality tree fruit harvester part II: Apples. Appl Eng Agric 19(5):545. https://doi.org/10.13031/2013.15314
24. Munic JP, Vougioukas SG, Arikapudi R (2016) A study on intercepting falling fruits with canopy penetrating rods. ASABE Paper No. 162456923, ASABE, St. Joseph, MI. https://doi.org/10.13031/aim.20162456923
25. Bulanon DM, Kataoka T, Zhang S, Ota Y, Hiroma T (1998) Optimal thresholding for the automatic recognition of apple fruits. ASAE Paper No. 013133. ASAE, St. Joseph, MI. https://doi.org/10.13031/2013.3672
26. Sarig Y (1993) Robotics of fruit harvesting: A state-of-the-art review. J Agric Eng Res 54(4):265–280. https://doi.org/10.1006/jaer.1993.1020
27. Peterson DL, Bennedsen BS, Anger WC, Wolford SD (1999) A systems approach to robotic bulk harvesting of apples. Trans ASAE 42(4):871. https://doi.org/10.13031/2013.13266
28. Silwal A, Karkee M, Zhang Q (2016) A hierarchical approach to apple identification for robotic harvesting. Trans ASABE 59(5):1079–1086. https://doi.org/10.13031/trans.59.11619
29. Zhao D, Lv J, Ji W, Ying Z, Yu C (2011) Design and control of an apple harvesting robot. Biosys Eng 110(2):112–122. https://doi.org/10.1016/j.biosystemseng.2011.07.005
30. Schupp J, Baugher T, Winzeler E, Schupp M, Messner W (2011) Preliminary results with a vacuum assisted harvest system for apples. Fruit Notes 76(4):1–5
31. Seavert CF, Freeborn J, Castagnoli S (2007) Orchard economics: Establishing and producing medium-density apples in hood river county. Or State Univ Extension Serv Bull. EM, 8829. Retrieved via https://agsci.oregonstate.edu/sites/agscid7/files/em8829-e.pdf on May 21. 2021
32. ASABE (2011a) D497.7: Agricultural machinery management data. ASABE, St. Joseph, MI
33. ASABE (2011b) EP496.3: Agricultural machinery management. ASABE, St. Joseph, MI
34. Edwards W (2015) Estimating farm machinery costs. Iowa State University Extension, Ames IA. Retrieved from https://www.extension.iastate.edu/agdm/crops/pdf/a3-29.pdf
35. Kay RD, Edwards WM, Duffy PA (2004) Farm management, 5th edn. McGraw-Hill, New York, NY
36. Mizushima A, Lu R (2011) Cost benefits analysis of in-field presorting for the apple industry. Appl Eng Agric 27(1):33–40. https://doi.org/10.13031/2013.29638
37. Zhang Z, Pothula A, Lu R (2016c) Economic analysis of a self-propelled apple harvest and infield sorting machine for the apple industry. ASABE Paper No. 2456644. ASABE, St. Joseph, MI. https://doi.org/10.13031/aim.20162456644

38. AgChoice Farm Credit (2021) Retrieved from https://www.agchoice.com/
39. U.S. Energy Information Administration (2021) Retrieved from https://www.eia.gov/electricity/state/
40. Fulmer S, Punnett L, Tucker Slingerland D, Earle-Richardson G (2002) Ergonomic exposures in apple harvesting: Preliminary observations. Am J Ind Med 42(S2):3–9. https://doi.org/10.1002/ajim.10087
41. Earle-Richardson GB, Fulmer S, Jenkins P, Mason C, Bresee C, May J (2004) Ergonomic analysis of New York apple harvest work using a Posture-Activities-Tools-Handling (PATH) work sampling approach. J Agric Saf Health 10(3):163–176. https://doi.org/10.13031/2013.16473
42. Brower MA, Earle-Richardson GB, May JJ, Jenkins PL (2009) Occupational injury and treatment patterns of migrant and seasonal farmworkers. J Agromedicine 14(2):172–178. https://doi.org/10.1080/10599240902799715
43. Baugher T, Schupp J, Lesser KM, Harsh RM, Lewis KM, Seavert CF, Auvil T (2009) Mobile platforms increase orchard management efficiency and profitability. Acta Hortic 824:361–364. https://doi.org/10.17660/actahortic.2009.824.42
44. Lu R, Zhang Z, Pothula AK (2017) Innovative technology for apple harvest and in-field sorting. Fruit Qtly 25(2):11–14
45. Sazo MM, De Marree A, Robinson T (2010) The platform factor-labor positioning machines producing good results for NY apple industry. New York Fruit Q 18(2):5–10
46. Gallardo RK, Galinato SP (2012) 2012 cost estimates of establishing, producing, and packing Red Delicious apples in Washington. Washington State University Extension. Retrieved from https://pubs.extension.wsu.edu/2012-cost-estimates-of-establishing-producing-and-packing-red-delicious-apples-in-washington

Chapter 4
Ergonomic Analysis of a Low-Cost Fresh Market Apple Harvest-Assist Unit

Zhaohua Zhang, Y. Qiao, H. Liu, Zhao Zhang, and M. Li

Abstract Manual apple harvest by seasonal migrant workers throughout the U.S. using ladder-bucket method. Though it is known that apple harvest would cause occupational injuries to workers, few studies have been conducted to quantify the levels of occupational injuries. Applying the rapid upper limb assessment (RULA) method, this study comprehensively evaluated the harvest process by dividing it into different activities. To alleviate occupational injuries in apple harvest, a low-cost harvest-assist unit has been developed, and its performance on reducing occupational injuries was further evaluated. Experimental results showed that seven out of 11 activities in the ladder-bucket harvest method would lead to occupational injuries, and the adoption of ladders and buckets was mainly responsible for causing occupational diseases. It has been demonstrated that adoption of the harvest-assist unit for high level apples harvest would not result in occupational injuries. Using the combined

Z. Zhang · M. Li (✉)
College of Economics and Management, Shandong Agricultural University, Tai'an 271018, China
e-mail: limin1988@163.com

Z. Zhang
e-mail: zzhang@sdau.edu.cn

Y. Qiao
Australian Centre for Field Robotics (ACFR), Faculty of Engineering, The University of Sydney, Sydney, NSW 2006, Australia
e-mail: yongliang.qiao@sydney.edu.au

H. Liu
Northwest Institute of Eco-Environment and Resources, Chinese Academy of Sciences, Lanzhou 730000, China
e-mail: lhayz@lzb.ac.cn

Z. Zhang
Key Lab of Modern Precision Agriculture System Integration Research, Ministry of Education of China, China Agricultural University, Beijing 100083, China
e-mail: zhang.zhao.1@ndsu.edu

Key Lab of Agriculture Information Acquisition Technology, Ministry of Agriculture and Rural Affairs of China, China Agricultural University, Beijing 100083, China

Department of Agricultural and Biosystems Engineering, North Dakota, State University, Fargo, ND 58102, USA

Z. Zhang et al. (eds.), *Mechanical Harvest of Fresh Market Apples*, Smart Agriculture 1,
https://doi.org/10.1007/978-981-16-5316-2_4

method (conventional for low level apples + harvest-assist unit for high level apples) to harvest apples would significantly reduce the occupational injury potential by cutting down the uncomfortable time ratio from 64 to 29%. This study demonstrated that combining the conventional harvest method for low level apples and the harvest-assist unit for high level apples harvest would significantly reduce the potential of occupational injuries. Thus, the newly developed apple harvest-assist unit supplies a solution to apple growers to alleviate harvest employees' occupational injuries.

Keywords Agricultural safety · Fresh market apple · Harvest-assist · Low-cost · Occupational injuries · Seasonal workers · Ergonomic analysis

4.1 Introduction

Apple is one of the most popular fruits globally [1]. Following China, U.S. is the second largest apple production country in the world [2]. With apples commercially grown in more than 32 states by more than 7500 growers, the farm-gate revenue of the U.S. apple crop is more than $ 5 $\times$ 10^9. Furthermore, the downstream economic activities related to apples are more than $ 15 $\times$ 10^9 [3]. In addition to fresh eating, apples can be consumed in a number of other manners, such as canned fruit and apple sauce [4, 5]. A summarized U.S. apple industry for the top five production states (Washington, New York, Michigan, Pennsylvania, and California) is shown in Table 4.1, and these top five states account for more than 95% of the total U.S. apple production. Washington is the leading apple production state, accounting for about 60% of the total production.

Though a lot of sensing and automation technologies have been developed and applied in agriculture, fresh market apples are still manually harvested using the conventional ladder-bucket approach, due to their susceptibility to bruising and lack of proper technologies [7–12]. For the low level apples, workers stand on the ground to pick, while for the high level apples, workers take advantage of ladders to reach and pick apples [13]. During the entire harvest process, workers wear a bucket to hold apples temporarily, and walk to the bin to release the harvested apples when the bucket is full [14]. Due to low efficiency of the ladder-bucket approach is inefficient, more than 50,000 seasonal migrant workers are required to complete the harvest process

Table 4.1 Summary data of U.S. apple industry [6]

State	Orchard number	Production area (ha)	Average yield (MT/ha)	Total yield ($\times$ 10^6 MT)	Fresh market apple (%)
Washington	2521	59,896	44.7	2.7	87.0
New York	1066	16,188	39.6	0.6	53.4
Michigan	1299	14,772	38.7	0.6	50.0
Pennsylvania	1239	8094	26.3	0.2	30.0
California	1984	6151	20.0	0.1	45.0

throughout the U.S. [15, 16]. In addition to inefficiency, the ladder-bucket approach is prone to causing occupational injuries [2]. On one hand, while standing on the ladder to reach apples, workers always stretch their bodies (awkward posture) to expedite the harvest process, as they are paid on a piece rate. On the other hand, wearing a bucket could also incur occupational injuries. When the bucket is full, it weighs about 20 kg, and workers need to wear such a heavy bucket to frequently conduct awkward activities, such as bending to release apples into the bin and climbing ladders [17–19].

After identifying the use of ladders and buckets is the major reason for occupational injuries, researchers from The Pennsylvania State University developed a low-cost harvest-assist unit to eliminate the use of ladders and buckets (refer to the Materials and Methods section). Working with the platform, workers only need to pick apples, and all other activates (e.g., apple transportation and bin filling) are completed automatically by the developed device. In addition to avoiding the occupational injuries, the use of platform could prevent the ladder fall accidents, which would easily result in fractures and even death, by installing tall protection rails [20].

Awkward postures have been demonstrated to be correlated to occupational injuries [2]. For the high level apples, workers need to conduct overhead work (e.g., arm and trunk extension and head tilting back) to get access to apples, which would cause back and shoulder disorders [21]. To identify the correlation between body postures and occupational injuries, posture-activity-tool-handling (PATH) method was widely employed in existing ergonomic studies [19]. The PATH method focuses more on the activity frequency and the general postures of the body, which provides little information on the body parts that are easy to be injured (e.g., neck, back, wrist and shoulder). Compared to the PATH approach, the rapid upper limb assessment (RULA) method has been developed with more factors (e.g., neck, back, leg, upper/lower limb) incorporated, which can provide a more detailed evaluation [22]. The reliability of the RULA method for evaluating occupational injuries has been validated in many studies [23–26]. Detached from existing literature, this study applied the RULA method to evaluate occupational injuries during apple harvest process, about which relative studies are scarce.

Various activities involved in ladder-bucket apple harvest process could lead to occupational injuries, such as wearing heavy buckets, and picking and dumpling apples. However, few studies have been conducted to quantify each activity's ergonomic performance and determine to what extent it would cause occupational injuries. Furthermore, though the developed apple harvest-assist unit could eliminate the use of ladders and buckets has been developed, its performance in reducing occupational injuries is unknown. Thus, the objectives of this study were to: (1) categorize the apple harvest process into different activities for three harvest methods: conventional ladder-bucket, harvest-assist unit, and the combination (conventional + harvest-assist unit); (2) calculate the RULA grand score for each activity and identify activities causing occupational injuries; and (3) compare the ergonomic evaluation results between different harvest methods.

Fig. 4.1 **a** Orsi Eco-pick mobile platform, and **b** harvest device mounted onto the platform

4.2 Materials and Methods

4.2.1 General Introduction to the Low-Cost Fresh Market Apple Harvest-Assist Unit

The low-cost harvest-assist unit consists of the mobile platform and the harvest device. To expedite the research progress, an off-the-shelf mobile platform (Eco-pick, Orsi, Group, Bologna, Italy) was purchased. The platform is powered by 12 V DC batteries, and can be used in the rugged field terrain. The harvest device replaces workers to transport apples into the bin, and it consists of receiver, tube, manifold, and distributor. Two workers stand on the platform to pick apples, and then put the harvested apples into the receiver. The tube, which is connected to the receiver, then transports apples downwards by gravity to the manifold. The manifold reduces apple velocity and then changes the fruit moving direction from horizontal to vertical, after which apples are delivered to the bin by the distributor. With the height of apples in the bin increase, the harvest device would be lifted gradually, in which way the automatic filling is realized. While working on the unit, the use of both ladders and buckets is avoided. Detailed information of the low-cost fresh market apple harvest-assist unit has been described by Zhang [14] and Zhang et al. [4, 5, 10, 16, 27], Zhang and Heinemann [28] (Fig. 4.1).

4.2.2 Rapid Upper Limb Assessment Method and Rating Procedure

RULA is a subjective assessment tool on body posture, which considers both the upper limb and lower body (Fig. 4.2; [29]). Since RULA method can only assess one posture at a time and apple harvest process is a motion state, it needs to break the

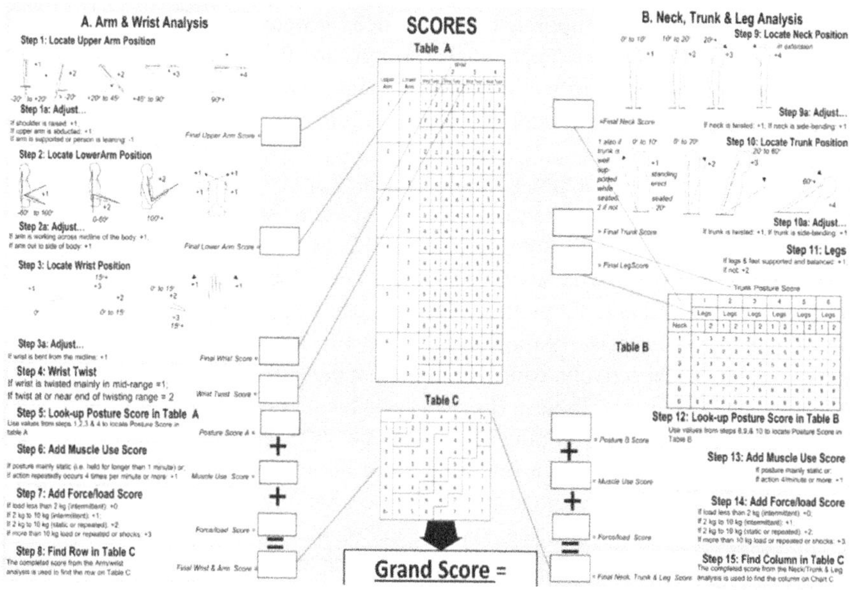

Fig. 4.2 Rapid upper limb assessment worksheet (copyright Alan Hedge, Cornell University, 2001) [29]

video into individual frames. For each individual frame, a RULA grand score would be obtained, and the grand score for a certain activity was calculated by averaging the scores of all frames. The grand score for RULA consists of two separate parts: one score assessing the arm/wrist postures, and the other one evaluating the neck/trunk/leg postures. Combining the two scores following Table C in Fig. 4.2, the final grand score would be obtained.

For the grand score, the larger the value, the more chances the activity would lead to occupational injuries. An activity with a RULA grand score 7 would be considered to link with serious occupational injuries; while an activity with a RULA grand score 1 would be considered to cause minimal occupational injuries. RULA score 5 is regarded as a threshold: if an evaluated posture has a RULA grand score > 5, it would cause occupational injuries; if an evaluated posture has a RULA grand score < 5, it is believed not to cause occupational injuries (Table 4.2; [2]).

Table 4.2 Rapid upper limb assessment (RULA) grades and indications [29]

RULA grand score	Action Level	Indications
1 or 2	1	Posture is acceptable
3 or 4	2	Further investigation is needed, and changes may be needed
5 or 6	3	Investigation and changes are required soon
7 or more	4	Investigation and changes are required immediately

Under the ladder-bucket approach, workers need to wear a bucket to hold harvested apples temporarily. Since an empty bucket weighs about 3 kg, it needs to be considered as a static load on the trunk. The static load increases with the number of apples put into the bucket. When 40 apples are stored into the bucket, the load comes to 10 kg (Step 7 in Fig. 4.2).

Releasing apples into the bin is a high frequency activity during the apple harvest process, and it is used as a case example to illustrate how to use RULA method for posture evaluation (Fig. 4.3). For step 1 in Fig. 4.2, the angle between trunk and arm is 65°, and the score is 3, which is adjusted to be 4 because the upper arm is abducted. For the step 2, the score is 2 as the lower arm has a 30° angle to the vertical line. For steps 3 and 4, the wrist position and its twist are reasonably assumed as 15° and mid-range twist, respectively, which are based on the authors' experience. Based on these obtained values and Table A (Fig. 4.2), a value of 4 is obtained as posture score A. Considering the apple dumping activity frequency is less than 4 times min^{-1}, the muscle use score is 1 (step 6 in Fig. 4.2). Since the total weight (apples + bucket) is more than 10 kg, the force/load score is 3 (step 7 in Fig. 4.2). By adding all these scores, the final wrist and arm score (step 8 in Fig. 4.2) is calculated as 8.

After obtaining the final wrist and arm score, the trunk posture score needs to be calculated. Based on the information provided in Fig. 4.3 and steps 9–11 in Fig. 4.2, the final neck score, final trunk score, and final leg score are obtained as 2, 4, and 1, respectively. Then, the posture B score is calculated as 5. After incorporating a muscle use score of 1 and a force/load score of 3, the final neck, trunk, and leg score (step 15 in Fig. 4.2) is obtained as 9. The RULA grand score 7 is obtained using Table C (Fig. 4.2), and the score 7 shows that this posture would cause seriously occupational injuries.

To increase harvest efficiency, workers usually use two arms for apple picking. Though two arms may have different postures, preliminary studies have demonstrated that using either arm for evaluation would not significantly affect the RULA grand score. Thus, during the posture evaluation process, the arm with less view block

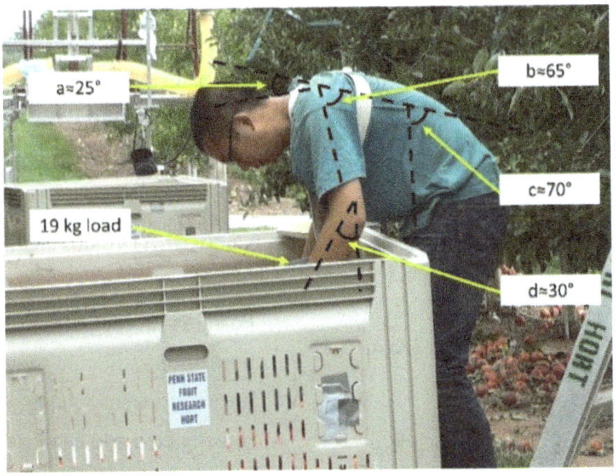

Fig. 4.3 Apple dumping posture: **a**, the angel between trunk and neck is 25°; **b**, the angle between trunk and upper arm is 65°; **c**, the angle between trunk and vertical line is 70°; **d**, the angle between lower arm and vertical line is 30°; the bucket is fully filled with apples and weighs 19 kg

would be selected. To avoid intra-rater difference, a single person, who has been trained by an ergonomic analysis specialist, rated all the frames.

In this study, all experiments were conducted at the Pennsylvania State Fruit Research and Extension Center (Adams County, PA) during the harvest window. Apple trees used in this study were about 3.5 to 4.0 m height, and structured in trellis wall format. A camcorder (HDR-CX550V 64 GB High-Definition Handycam, Sony Corp., Tokyo, Japan) was used for recording the harvest process for different experiments.

4.2.3 Ladder-Bucket Harvest Approach

4.2.3.1 Video Recording

Four workers used the ladder-bucket approach to harvest 15 'Fuji' apple trees (five trees as a group for three replications). While video recording the harvest process, the cameraman needed to find a good view to expose as much as the worker's body with minimal view block. For all the trees, workers only harvest 50% of total apples (only picking those facing them), as the trellis wires blocked workers from getting access to the other half apples on the canopy.

Since the RULA method can only assess postures, and the collected data are in video format, individual frames were extracted from the video. Since no single activity could be completed in less than 0.5 s, frames were extracted from the video at a frequency of 2 frames per second (fps) using Photoshop (Adobe Systems, San Jose, CA, U.S.A.).

4.2.3.2 Activity Categorization and Ergonomic Evaluation

The relative location between apples and workers would affect the results of ergonomic analysis. When a worker stands on the ground neutrally with a bucket worn, the high, middle and low level apples are defined as apples higher than the shoulders, between the bottom of the bucket and the shoulder, and below the bucket bottom, respectively. While a worker stands on the ladder for picking, the same principles are applied for defining high, middle, and low level apples. Then, the conventional apple harvest process is categorized into 12 different activities: picking low, middle, and high level apples on the ground and ladder, moving/climbing/descending ladder, walking to and away from bin, and dumping apples (Fig. 4.4). The corresponding frames were then evaluated to get a grand score for each frame, with the average used to represent the activity's grand score.

Fig. 4.4 Categorized apple harvest activities using conventional ladder-bucket method: I, II, and III are picking low, middle, and high level apples on the ground, respectively; IV is moving ladder; V, VI, and VII are picking middle, high, and low level apples on the ladder, respectively; VIII is climbing ladder; IX, X, XI, and XII are descending ladder, waling to bin, dumping apples, and walking away from bin, respectively

4.2.4 *Low-Cost Harvest-Assist Unit for Apple Harvesting*

4.2.4.1 Video Recording

The worker stood on the platform to harvest the high level apples of 15 'Fuji' apple trees, and a cameraman video recorded the whole process. The worker only harvested 50% of the apples in the tree: the side facing the worker. For the 15 trees, every 5 trees were considered as a replication, and thus there were totally 3 replications. To have a better view of the workers working on the platform, the cameraman stood on a ladder to take the video to minimize the view block on the worker.

4.2.4.2 Activity Categorization and Ergonomic Evaluation

Different from the conventional ladder-bucket harvest method, activities using the harvest-assist unit are relatively fewer, and only include picking low, middle, and high level apples (Fig. 4.5). After completing one tree harvest, the worker drove the unit forward to the next apple tree. Since the driving time was very short (1~2 s) and the activity of driving the unit would not cause occupational injuries, the unit driving activity was not included in this study.

A B C

Fig. 4.5 Categorized apple harvest activities using a low-cost harvest-assist unit: A, B, and C are picking middle, low, and high level apples

4.2.5 Combined Method (Conventional + harvest-Assist Unit)

4.2.5.1 Video Recording

While applying the harvest-assist unit for apple picking, only the high level apples would be harvested. For those low level apples, workers need to take advantage of the conventional method to complete the harvest. However, since the high level apples are harvested by the harvest-assist unit, there is no need to use the ladders while applying the conventional method. A set of 15 'Fuji' apple trees were harvested using the combined method by the same worker. The worker first used the harvest-assist unit to harvest the high level apples, and then stood on the ground by wearing a bucket to pick the low level apples. The 15 trees were also divided into 3 replications, with every 5 neighboring trees as a replication. The cameraman paid more attention to avoid view block on the worker during video recording.

4.2.5.2 Activity Categorization and Ergonomic Evaluation

The activities of using the combined method are list in Table 4.3. There are a total of nine activities, including three by using the harvest-assist unit on picking high level apples (Table 4.3, *1*, *2*, *3*), and six for picking low level apples on the ground.

4.3 Results and Discussion

4.3.1 Conventional Ladder-Bucket Harvest Approach

A 25 min video was recorded, and then a total of 3000 frames were assessed, with the evaluation results show in Fig. 4.6. Among the 11 activities, only four activities (yellow color bars in Fig. 4.6) are identified as not to cause occupational injuries, and the other seven activities would lead to occupational diseases. For the activity of walking away from the bin, since the bucket is empty (all apples released into the bin)

Table 4.3 Categorized activities using combined method (conventional + harvest-assist unit)

Activity number	Activity
1	Picking low level apples on the harvest-assist unit
2	Picking middle level apples on the harvest-assist unit
3	Picking high level apples on the harvest-assist unit
4	Picking low level apples on the ground
5	Picking middle level apples on the ground
6	Picking high level apples on the ground
7	Walking to bin
8	Walking away from bin
9	Releasing apples

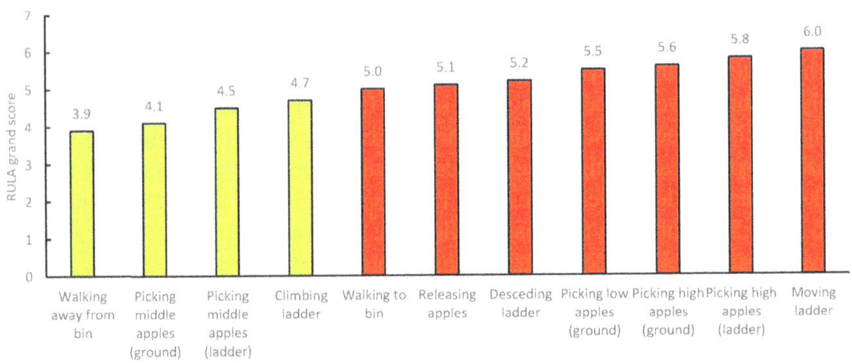

Fig. 4.6 RULA grand scores of different activities under the conventional ladder-bucket apple harvest approach. Activities with RULA grand score >5 are considered as resulting in occupational injuries. RULA: rapid upper limb assessment

and the worker walks back to the tree comfortably, it is reasonable that the RULA grand score is low. For activities of picking middle apples on the ground or ladder, since the worker does not need to stretch his body to get access to high or low level apples, the picking postures are not awkward, which explains the small values of RULA grand scores. This could also explain why the RULA grand scores are above 5 for activities of picking low and high apples on the ground and picking high apples on ladder. The activity of climbing ladder demonstrates not to incur occupational injuries is mainly due to the bucket is not full. However, the descending ladder activity has a RULA grand score higher than 5, which is because the bucket is full while the worker climbs down the ladder. The walking to bin activity is risky, because walking while wearing a 20 kg fully filled bucket would easily lead to occupational injuries to the trunk and back. The risky activity of releasing apples is also caused by

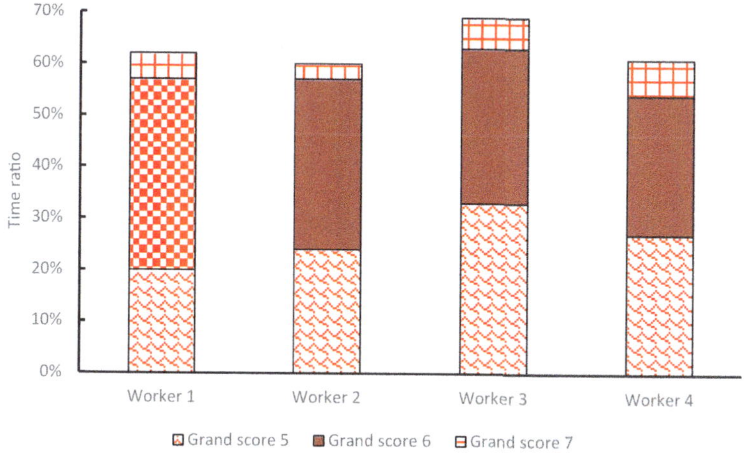

Fig. 4.7 Time ratios of four different workers spent on activities causing occupational injuries

the fully filled heavy bucket (~19 kg). As walking is more comfortable than bending, the RULA grand score for walking to the bin is a little smaller than that of releasing apples. The most dangerous activity is moving ladder, which has the largest RULA grand score (6.0). The ladder itself is heavy, ranging from 10 to 20 kg. In addition, the orchard terrain is uneven, and walking in an uneven field by holding a heavy ladder is easily to result in sprains or strains.

After identifying activities that are easy to cause occupational injuries, this study also analyzed the time percentages of the workers spent on these activities, which is shown in Fig. 4.7. For the four workers, their time ratios spent on occupational disease-prone activities (RULA grand score > 5) ranges from 60 to 70%, with an average of 64%. In another word, workers spend two thirds of their working time on activities that would cause occupational injuries, which quantitatively demonstrated that conventional apple harvest process needs improvement.

It can be further noticed that a majority of the activities that lead to occupational injuries are due to the use of ladders and buckets. Thus, the adoption of the harvest-assist unit would help to alleviate occupational injuries by replacing ladders with the unit and eliminating the use of buckets.

4.3.2 Harvest Platform Evaluation Results

After recorded a 11 min video, about 1300 frames were extracted and assessed. The averaged grand score for each activity is shown in Fig. 4.8. All the three activities have a grand score smaller than 5, indicating all of them would not cause occupational injuries. Compared to the conventional ladder-bucket harvest method consisting of a lot of activities causing occupational injuries, working on the harvest-assist unit

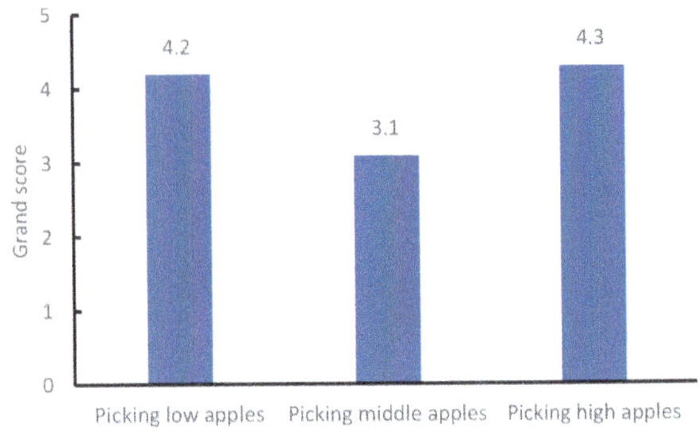

Fig. 4.8 Rapid upper limb assessment grand scores for three activities using harvest-assist unit

would not result in health issues. The activity of picking middle level apples has the lowest RULA grand score because the worker does not need stretch his body to get access to apples. However, when picking low and high level apples, workers need to frequently bend or stretch their bodies to reach apples, generating high RULA grand scores.

4.3.3 Combined Method Evaluation Results

A 18 min video was recorded for the combined harvest method, which resulted into 2200 frames. After categorizing the activity and conducting RULA evaluation, the results are shown in Table 4.4. In the combined harvest method, workers standing on the platform only picked middle and high level apples because picking low level apples needed workers to bend their bodies significantly to get access to the apples, which is abandoned by the workers. Two activities get a RULA grand score larger than

Table 4.4 Rapid upper limb assessment (RULA) evaluation results of different activities and their corresponding time percentages under combined harvest method (conventional + harvest-assist unit)

Activity	RULA grand score	Time ratio
Picking low level apples on the harvest-assist unit	0	0
Picking middle level apples on the harvest-assist unit	3.9 ± 0.3	37%
Picking high level apples on the harvest-assist unit	4.4 ± 0.5	2%
Picking low level apples on the ground	5.9 ± 0.8	10%
Picking middle level apples on the ground	4.4 ± 0.5	29%
Picking high level apples on the ground	5.5 ± 0.6	12%
Walking to bin	5.1 ± 0.3	4%
Walking away from bin	4.7 ± 0.3	2%
Releasing apples	5.6 ± 0.8	3%

5—picking low and high level apples on the ground. This is because workers have to stretch or bend to complete apple picking. Based on the time ratios from Table 4.4, activities with RULA grand score >5 accounts for 29% of the total working time, which is much less than that of the conventional ladder-bucket method of 64% (Fig. 4.7). These results indicate that the adoption of the harvest-assist unit would alleviate the occupational injuries.

4.4 Conclusion

Ergonomic analysis for three different harvest methods were conducted including conventional ladder-bucket, harvest assist-unit, and the combination of the two methods. For the conventional method, workers spent about two thirds of their total working time on activities with a rapid upper limb assessment (RULA) grand score >5, which is easy to lead to occupational injuries. Furthermore, it demonstrates the use of ladders and buckets are the two major causes of occupational injuries. Evaluation of applying the low-cost harvest-assist unit showed that the RULA grand scores for all activities are lower than 5, indicating the application of the harvest-assist unit would not result in occupational injuries. This confirms that the harvest-assist concept by eliminating the use of ladders and buckets does reduce the occurrence of occupational injuries. Experimental results of the combination harvest methods—harvest-assist unit for the high level apples and conventional method on the middle and low level apples—show that the combined method significantly reduces the uncomfortable working time ratio from 64% of the conventional method to 29% of the combined method. Therefore, workers' occupational injuries would be significantly reduced by the adoption of the combined method.

Since this study only focused on orchards in Pennsylvania state, further studies should consider the orchard variance in different states to demonstrate the technology reliability and robustness of the findings. Furthermore, in current study, workers manually/visually evaluate the RULA grand score for each posture/frame, which is inefficient and subjective. With the progress of machine vision, imaging process, and advanced machine learning algorithms, an automated method should be developed.

Credit Authorship Contribution Statement **Z. Zhang**: Writing—original draft. **Y. Qiao**: Writing—review & editing. **H. Liu**: Writing—review & editing. **Z. Zhang**: Conceptualization, Investigation, Methodology, Project administration, Writing—original draft, Writing—review & editing. **M. Li**: Conceptualization, Investigation, Methodology, Project administration, Writing—original draft, Writing—review & editing, Writing—review & editing.

Disclaimer
Mention of commercial products or orchards in this paper is only for providing factual information and does not imply endorsement of them by authors over those not mentioned.

Declaration of Competing Interest
The authors declare that they have no known competing financial interests or personal relationships that could have appeared to influence the work reported in this paper.

References

1. Lu Y, Lu R (2016) Quality evaluation of apples. In: Sun D-W (ed) Computer vision technology for food quality evaluation, 2nd edn. Academic Press, San Diego, CA, pp 273–304. https://doi.org/10.1016/B978-0-12-802232-0.00011-6
2. Zhang Z, Zhang Z, Wang W, Liu H, Sun Z (2019) The role of a new harvest platform in alleviation of apple workers' occupational injuries during harvest. J Agric Saf Health 25(1):11–24. https://doi.org/10.13031/jash.13103
3. U.S. Apple Association (2021) Industry at a glance. Retrieved from https://usapple.org/
4. Zhang Z, Heinemann PH, Liu J, Schupp JR, Baugher TA (2017a). Brush mechanism for distributing apples in a low-cost apple harvest-assist unit. Appl Eng Agric 33(2):195–201. https://doi.org/10.13031/aea.11971
5. Zhang Z, Pothula AK, Lu R (2017b) Development of a new bin filler for apple harvesting and infield sorting with a review of existing technologies. ASABE Paper No. 1700662. ASABE, St. Joseph, MI. https://doi.org/10.13031/aim.201700662
6. USDA (2015) Noncitrus fruits and nuts: 2014 summary. USDA National Agricultural Statistics Service, Washington, DC. Retrieved from https://downloads.usda.library.cornell.edu/usdaesmis/files/zs25x846c/6108vd86r/3r074x570/NoncFruiNu-07-17-2015.pdf
7. Flores P, Zhang Z, Igathinathane C, Jithin M, Naik D, Stenger J, Ransom J, Kiran R (2021) Distinguishing seedling volunteer corn from soybean through greenhouse color, color-infrared, and fused images using machine and deep learning. Ind Crops Prod 161:113223. https://doi.org/10.1016/j.indcrop.2020.113223
8. Jahan N, Flores P, Liu Z, Friskop A, Mathew JJ, Zhang Z (2020) Detecting and distinguishing wheat diseases using image processing and machine learning algorithms. ASABE Paper No. 2000372. ASABE, St. Joseph, MI. https://doi.org/10.13031/aim.202000372
9. Yao L, Hu D, Zhao C, Yang Z, Zhang Z (2021) Wireless positioning and path tracking for a mobile platform in greenhouse. Int J Agric Biol Eng 14(1):216–223. https://doi.org/10.25165/j.ijabe.20211401.5627
10. Zhang Z, Heinemann PH, Liu J, Baugher TA, Schupp JR (2016a) The development of mechanical apple harvesting technology: a review. Trans ASABE 59(5):1165–1180. https://doi.org/10.13031/trans.59.11737
11. Zhang Z, Igathinathane C, Li J, Cen H, Lu Y, Flores P (2020) Technology progress in mechanical harvest of fresh market apples. Comput Electron Agric 175:105606. https://doi.org/10.1016/j.compag.2020.105606
12. Zhang Z, Lu Y, Lu R (2021) Development and evaluation of an apple infield grading and sorting system. Postharvest Biol Technol 180:111588. https://doi.org/10.1016/j.postharvbio.2021.111588
13. Freivalds A, Park S, Lee C, Earle-Richardson G, Mason C, May JJ (2006) Effect of belt/bucket interface in apple harvesting. Int J Ind Ergon 36(11):1005–1010. https://doi.org/10.1016/j.ergon.2006.08.005
14. Zhang Z (2015) Design, test, and improvement of a low-cost apple harvest-assist unit. Ph.D. dissertation. University Park, PA: Pennsylvania State University, Department of Agricultural and Biological Engineering
15. Luo R, Lewis KM, Zhang Q, Wang SM (2012) Assessment of bruise damage by vacuum apple harvester using an impact recording device. ASABE Paper No. 121338094. ASABE, St. Joseph, MI. https://doi.org/10.13031/2013.41870
16. Zhang Z, Heinemann P, Liu J, Schupp J, Baugher T (2014) Design, fabrication, and testing of a low-cost apple harvest-assist device. ASABE Paper No. 141839738. ASABE, St. Joseph, MI. https://doi.org/10.13031/aim.20141839738
17. Earle-Richardson G, Jenkins PL, Strogatz D, Bell EM, May JJ (2006) Development and initial assessment of objective fatigue measures for apple harvest work. Appl Ergon 37(6):719–727. https://doi.org/10.1016/j.apergo.2005.12.002

18. Earle-Richardson G, Jenkins P, Fulmer S, Mason C, Burdick P, May J (2005) An ergonomic intervention to reduce back strain among apple harvest workers in New York State. Appl Ergon 36(3):327–334. https://doi.org/10.1016/j.apergo.2004.12.003
19. Earle-Richardson GB, Fulmer S, Jenkins P, Mason C, Bresee C, May J (2004) Ergonomic analysis of New York apple harvest work using a Posture-Activities-Tools-Handling (PATH) work sampling approach. J Agric Saf Health 10(3):163–176. https://doi.org/10.13031/2013.16473
20. Fathallah FA (2010) Musculoskeletal disorders in labor-intensive agriculture. Appl Ergon 41(6):738–743. https://doi.org/10.1016/j.apergo.2010.03.003
21. Sakakibara H, Miyao M, Kondo T-A, Yamada S (1995) Overhead work and shoulder-neck pain in orchard farmers harvesting pears and apples. Ergonomics 38(4):700–706. https://doi.org/10.1080/00140139508925141
22. Bernard BP (1997) Musculoskeletal disorders and workplace factors: a critical review of epidemiologic evidence for work-related musculoskeletal disorders of the neck, upper extremity, and low back. NIOSH, Cincinnati, OH
23. Dockrell S, O'Grady E, Bennett K, Mullarkey C, Mc Connell R, Ruddy R, Twomey S, Flannery C (2012) An investigation of the reliability of Rapid Upper Limb Assessment (RULA) as a method of assessment of children's computing posture. Appl Ergon 43(3):632–636. https://doi.org/10.1016/j.apergo.2011.09.009
24. Drinkaus P, Sesek R, Bloswick D, Bernard T, Walton B, Joseph B, Reeve G, Counts JH (2003) Comparison of ergonomic risk assessment outputs from Rapid Upper Limb Assessment and the Strain Index for tasks in automotive assembly plants. Work 21(2):165–172
25. Fountain LJK (2003) Examining RULA's postural scoring system with selected physiological and psychophysiological measures. Int J Occup Saf Ergon 9(4):383–392. https://doi.org/10.1080/10803548.2003.11076576
26. Sharan D, Ajeesh PS (2012) Correlation of ergonomic risk factors with RULA in IT professionals from India. Work 41:512–515. https://doi.org/10.3233/wor-2012-0205-512
27. Zhang Z, Heinemann PH, Liu J, Schupp JR, Baugher TA (2016b) Design and field test of a low-cost apple harvest-assist unit. Trans ASABE 59(5):1149–1156. https://doi.org/10.13031/trans.59.11708
28. Zhang Z, Heinemann P (2017) Economic analysis of a low-cost apple harvest-assist unit. HortTechnology 27(2):240–247. https://doi.org/10.21273/horttech03548-16
29. McAtamney L, Corlett EN (1993) RULA: a survey method for the investigation of work related upper limb disorders. Appl Ergon 24(2):91–99. https://doi.org/10.1016/0003-6870(93)90080-s

Chapter 5
Development, Evaluation and Improvement of Apple Infield Grading and Sorting Systems

Zhao Zhang and Y. Lu

Abstract Infield grading and sorting could help apple growers achieve cost savings in postharvest handling by removing inferior fruit that are not suitable for fresh market consumption. An apple infield grading and sorting system, including subsystems of singulation and rotation, grading, and sorting, was developed. The grading system's repeatability coefficients >90% and >81% for the intra- and inter-lane, respectively, demonstrated its satisfactory performance for infield use. The rotary sorter (RS), consisting of a rotary disc and two pivotable gates to guide apples into different bins according to grading results, could not meet the requirements in terms of high apple bruising rate (<95% apples in Extra Fancy level), bulky sizes (610 mm height), complex hydraulic power transmission system, and unacceptably low sorting accuracy. To address the issues caused by RS, a simple and compact paddle sorter (PS), consisting of a solenoid and a paddle, was designed and constructed. Extensive tests of the PS resulted in 100% of apples graded as Extra Fancy, which had a superior performance over RS. By replacing the RS with the PS, the overall height of the sorting system reduced from 610 to 350 mm, making it more convenient for transportation and infield use. Furthermore, the hydraulic power transmission system was simplified significantly by eliminating a majority of components (e.g., angle gearbox, coupler, and sprocket) when the RS was replaced by the PS. The PS achieved sorting accuracy >99%, which demonstrated its satisfactory and reliable performance. The apple infield grading and sorting system with the PS incorporated

---

Z. Zhang (✉)
Key Lab of Modern Precision Agriculture System Integration Research, Ministry of Education of China, China Agricultural University, Beijing 100083, China
e-mail: zhang.zhao.1@ndsu.edu

Key Lab of Agriculture Information Acquisition Technology, Ministry of Agriculture and Rural Affairs of China, China Agricultural University, Beijing 100083, China

Department of Agricultural and Biosystems Engineering, North Dakota, State University, Fargo, ND 58102, USA

Y. Lu (✉)
Department of Agricultural and Biological Engineering, Mississippi State University, Starkville, MS 39762, USA
e-mail: yzlu@abe.msstate.edu

Z. Zhang et al. (eds.), *Mechanical Harvest of Fresh Market Apples*, Smart Agriculture 1,
https://doi.org/10.1007/978-981-16-5316-2_5

is bruising free, compact, simple, and satisfactory in sorting accuracies (>99%), indicating its potential for long-term commercial use.

Keywords Apple · Infield sorting · Grading · Automation · High throughput · Bruising

5.1 Introduction

Apple has been manually harvested using ladder-bucket methods for decades [1]. During the harvesting process, seasonal workers stand on the ground to pick low-level apples and take advantage of ladders to get access to high-level fruits [2–4]. Simultaneously, workers wear a bucket to temporarily hold apples during the entire harvest process, and when the bucket is full, they walk to a bin to release apples [5, 6]. Considering the short harvest window (around 1.5 months), tight labor pool, and piece-rate payment, seasonal workers are impatient and unable to conduct apple infield sorting [7]. Hence, apples of mixed quality grades are stored in the same bins until they are full, after which bins are hauled into packinghouses for storage [1, 8]. When orders from retailers are placed, apples would be pulled out of the storage room for grading, sorting, and packaging. If a majority of harvested apples are at fresh market level, growers earn good benefits, though the high cost occurring with the postharvest handling (i.e., storage, grading, sorting, and packaging). However, if processing (or even cull) apples account for a major part, growers may not break even by factoring in the high postharvest handling cost and low processing level apple price [9]. Hence, it is desirable to sort and then store apples into different bins according to their qualities infield.

If the same grade apples are stored in the same bins infield, they could be treated in a more economic manner during the postharvest handling: processing apples would be stored in cold storage (~\$30 MT^{-1} $month^{-1}$) and fresh market apples in controlled atmosphere (~\$80 MT^{-1} $month^{-1}$) (personal communication with Riveridge Packing LLC, and Elite Apple Co. LLC, MI, US). Furthermore, the high cost for grading, sorting, and packaging (~\$290 MT^{-1}) would exclusively occur to fresh market apples, in which way it would significantly reduce the cost on the processing apples. A previous case study confirmed that for 10 MT apples (6 MT fresh market and 4 MT processing), the infield sorting technology (90% sort-out accuracy) could save around 35% of the total postharvest handling cost [10]. Thus, infield sorting has been identified as one of the top 10 technologies needed by apple growers [11]. Though infield sorting technology's commercial potential has been thoroughly evaluated and validated by Mizushima and Lu [12] and Zhang et al. [10], there lacks the commercial products and few infield fruit grading and sorting projects have been reported, due to the demanding requirements of the technology (e.g., high throughput, reliability, and robustness).

Infield grading and sorting system includes sub-systems of apple singulation, rotation, grading, and sorting [7]. Fruit singulation is to separate overlapping apples

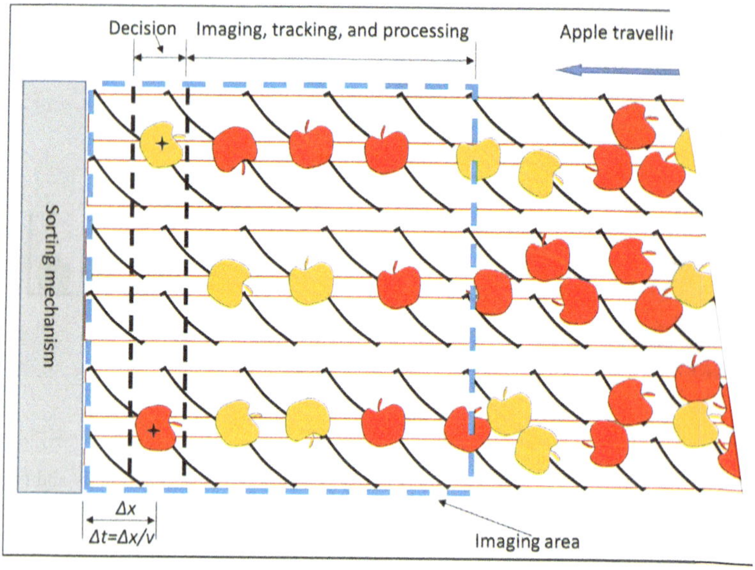

into an individual linear format [13]. An extensively-adopted separation method is to transport apples via multiple belts with different speeds [14]. When shifting from low- to high-speed belts, apples are physically separated due to the acceleration. Another singulation approach applies liquids medium for fruit conveyance, and uses a V-shaped cross-section conveyor and diabolo-shaped element for fruit separation [15]. Agrosaw [16] fed apples into a spinning disk and took advantage of centrifugal force to realize fruit singulation. However, all these existing fruit singulation technologies are adopted in packinghouses, and their bulky sizes and complexities prevent them from infield use.

Fruit rotation exposes the entire apple surface under the machine vision system for objective grading. Bi-cone rollers have been commercially used in packinghouses to drive fruit to rotate by friction [13, 17, 18]. Throop et al. [19] designed a rotation mechanism, including a ring and a wheel protruding up through the center of the ring. The ring holds an apple loosely and the rotating wheel drives apple's rotation when contacting the fruit cheek [20–22]. Additionally, Blasco et al. [23] developed a movable suction cup to hold and then rotate fruit for the presence of its full surface for camera inspection. However, none of them is suitable for long-term high throughput infield use: the bi-cone roller and rotating wheel may dysfunction when exposed in dusty and vibrant working conditions, and the suction cup is improper for high throughput use (6 to 9 apple s^{-1}).

Grading sub-system evaluates fruit quality in terms of size and color in real-time mode [24–26]. Since grading sub-system would be mounted above the fruit rotation mechanism, it requires the imaging chamber to be compact to limit the overall height [27]. Cubero et al. [28] fixed a camera parallel to the fruit moving direction and then mounted a tilted mirror above fruit to reflect fruit images to the camera. Though this innovative setup shortened the imaging chamber's height, it may not perform robustly for long-term use, due to dust and machine vibration.

The sorting sub-system guides apples into different destinations according to grading results [25, 29, 30]. A commercially adopted method is to hold and track apples in individual cups, and when one apple arrives at its final destination, the cup is triggered to be tilted and then the fruit exits the cup for its destination [31, 32]. In addition to the cup method, researchers [14, 33–35] employed air jet principle for apple sorting. Instead of triggering cups, this method triggered an air ejector to push apples to its destination. However, the complex mechanical system of cup design and random apple movement caused by the air jets prevent their infield application.

Towards the goal to develop an infield grading and sorting system, the objectives of this study were to: (1) evaluate the intra- and inter-lane grading repeatability of the machine vision-based grading system; (2) design and test a rotary sorter (RS) and identify the room for improvement, and (3) design and test a new paddle sorter (PS) in terms of bruising and sorting accuracy.

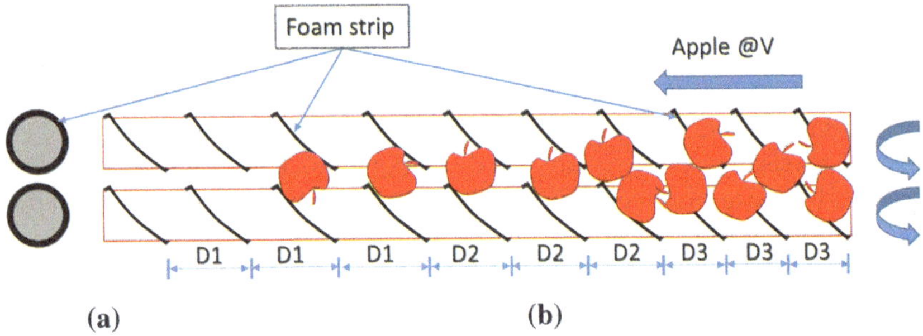

Fig. 5.1 A set of pitch-variable screw conveyors for apple singulation, rotation, and transportation: **a** side view and **b** top view. The pitches of D3 (83 mm) < D2 (102 mm) < D1 (121 mm)

5.2 Materials and Methods

5.2.1 Apple Singulation and Rotation Sub-system

A pitch-variable screw conveyor to realize fruit singulation, rotation, and transportation is developed (Fig. 5.1). While the screw conveyors rotate (synchronized in the same direction), the protruding soft foam strips (multi-purpose neoprene foam strips, McMaster-Carr, Aurora, OH, U.S.) attached to the surface of the conveyor push fruit forward for transportation. The pitch increases (D1 > D2 > D3 in Fig. 5.1) along the direction fruit moving forward, and while moving forward, apples in cluster are separated into tandem arrangements. While the screw conveyor rotates, the friction between fruit and the conveyors rotates apples in reverse direction to rotating conveyors, and when apples arrive at the end of screw conveyors, they are already separated. The full system consists of three sets of screw conveyors (total six synchronized conveyors spinning at the same direction) and was validated to be capable of handling maximal 12 apple s^{-1} (4 apple s^{-1} lane^{-1}).

5.2.2 Grading Sub-system

The grading sub-system consists of an imaging chamber and image processing algorithms [24, 36]. The imaging chamber (height 610 mm) is used to mount a low cost CCD color camera (Fire-I Digital Camera, Unibrain Inc., CA, U.S.) for image collection and eight lamps (T8 led lights, Forest Lighting Inc., Marietta, GA, U.S.) for uniform illumination. The imaging chamber is mounted at the end of screw conveyors, so when apples arrive at the chamber, they are already separated. A PCI-express adapter (Fireboard Blue-e 1394a, Unibrain Inc., CA, USA) is used as a medium between the camera and computer for real-time image collection, storage, and process.

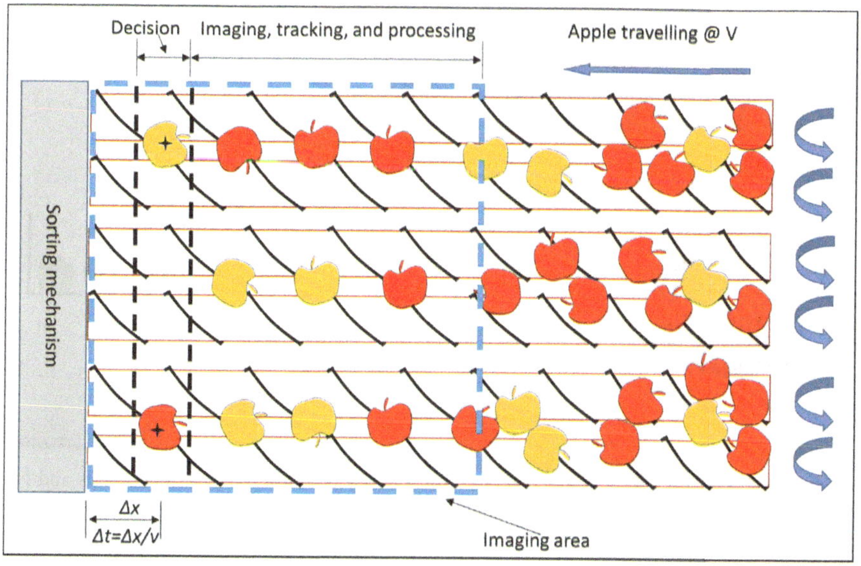

Fig. 5.2 Imaging area for fruit grading (i.e., fresh or processing) and tracking, and time estimation for apples to arrive at the sorting mechanism

The grading algorithms process collected images and then make a decision on apple grades. Since each collected image covers partial surface of an apple, the algorithm tracks an individual apple and applies the average of 5~10 image grading results as the fruit's final grade. The evaluation process is completed before the apple arrives at the decision zone (Fig. 5.2). After an apple fully enters the decision zone, the collected images would be used to determine the centroid of individual apples. Combining the distance (Δx) between apple centroid and sorting mechanism and the apple real-time speed (v; obtained from a speed sensor), the time needed for individual apples to arrive at the sorting mechanism is calculated. Based on the calculated time, the algorithm would send signals via DAQ device (USB-6009, National Instruments, Austin, TX, U.S.) to trigger the sorting system to guide apples into corresponding destinations. Tests of the machine vision system demonstrate it is capable of acquiring and processing one image (with multiple apples) within 50 ms (image capture, size estimation, and color assessment for approximately 12 ms, 20 ms, and 15 ms, respectively), and this meets the practical application requirements [7, 12, 24].

5.2.3 *Rotary Sorter*

The RS consists of a rotary dick and two gates to guide apples to different destinations according to the grading results (Fig. 5.3). The current RS has been divided into four

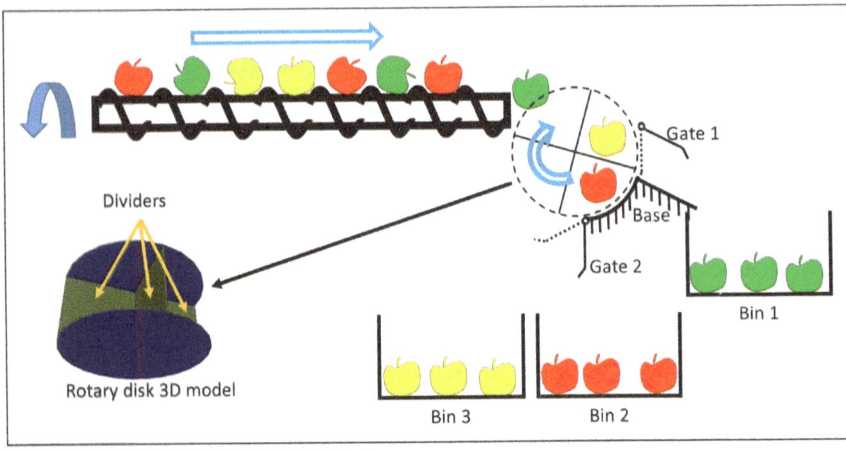

Fig. 5.3 A developed rotary sorter consisting of a rotary disk with four compartments and two gates for guiding apples to different destinations

compartments and is mounted immediately behind the end of pitch-variable screw conveyors. Since the screw conveyor and RS are synchronized, when the screw conveyor rotates four revolutions, the RS completes one revolution, which guarantees each apple would take one compartment. The grading results by the imaging subsystem determine the proper timing of open and close of different gates. If an apple is graded for Bin 1, the Gate 1 would be triggered to open when the apple arrives, and gravity and centrifugal force would release the apple out of the rotary disk via open Gate 1. For apples destined to Bin 2 or 3, the Gate 1 would be in close mode when the apples arrive, and the Gate 2 would be in open mode for an apple destined to Bin 2, and in close mode for an apple targeted to Bin 3.

After the RS was designed and fabricated, it was tested, with several issues being identified. When the RS works at high throughput, even the Gate 1 is open, an apple does not have enough time to fully roll out of the compartment and this could lead to three potential undesirable scenarios—misclassification, apple cut, and Gate 1 clogging. The misclassification refers to that apples that should be guided to Bin 1 finally arrive at Bin 2 or 3. When the time is not enough for an apple to be released from Gate 1 before the next divider arrives, the apple could be stuck between the divider and the base (Fig. 5.3). Due to the powerful hydraulic system (1/2 hp), the stuck apple is inevitably cut into two halves, and this is observed multiple times. The Gate 1 stuck issue is that while the apple rolling out of the disc, the gate closes and then the apple is stuck between the Gate 1 and Base. Since the Gate 1 is not powerful enough to cut apples, the stuck apple would keep the Gate 1 open constantly, which would affect the later arrived apples.

After the RS was completed, it was incorporated into the apple harvest and infield sorting (HIS) machine. A trailer was needed to haul the HIS machine from machine shop to an apple orchard for field tests. However, after loading the HIS machine onto a standard trailer for transportation, the overall height was 150 mm (6 in.) higher than

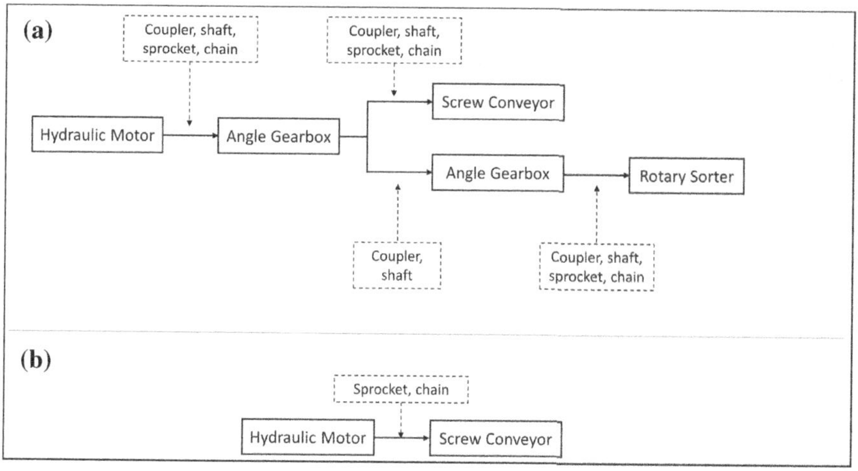

Fig. 5.4 Diagram of hydraulic and mechanical power transmission of rotary sorter (**a**) and paddle sorter (**b**)

the height limitation for road transportation. Thus, more efforts should be invested to shorten the RS height.

The RS had a complex hydraulic power transmission system (Fig. 5.4a), and it took a lot of space for mounting. The hydraulic motor power was transmitted to an angle gearbox (Andantex R3303) with two synchronized outputs. One output powered the three screw conveyor sets, and the other output powered the RS—another angle gearbox was needed to change the power direction. The complexity of power transmission system was mainly attributed to the requirement of synchronizing the screw conveyors and RS. It was also concerned of the complex hydraulic system's reliability for its long-term infield use.

5.2.4 *Paddle Sorter*

The bulky and complex RS was replaced by a simple paddle sorter (PS) design, consisting of a rotary solenoid (stroke rotation 40°, 16 oz. torque, 12 VDC; McMaster-Carr, Aurora, OH, U.S.) and a paddle. The PS has two positions as open and close, controlled by the grading results. When the PS is open (Fig. 5.5b; fresh market apples), it would not interfere the apple movement trajectory—the apple continues to move forward after exiting the end of screw conveyors. When the PS is in close mode (Fig. 5.5c), it would change the apple movement trajectory significantly by pushing the fruit to the other destination. By controlling the paddle open and close, apples of different grades are guided to corresponding destinations.

The PS design is superior over the RS in a number of aspects. First, it eliminates the apple cut issue, as there is no chance for apples to get stuck. Second, the overall

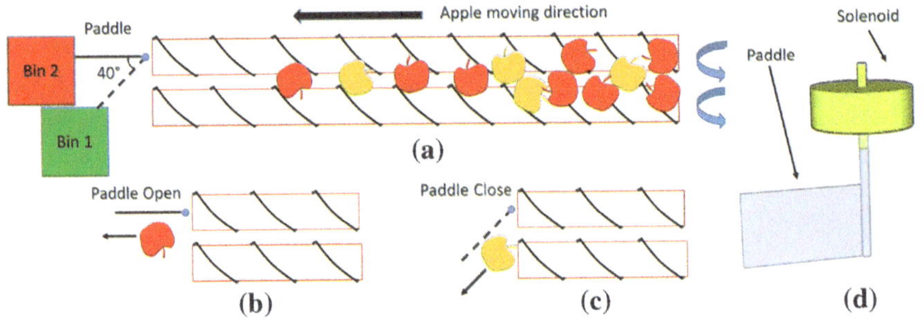

Fig. 5.5 Paddle sorter coupled with pitch-variable screw conveyors: **a** top view layout of paddle with screw conveyor, **b** fresh apple moving trajectory, **c** cull apple moving trajectory, and **d** paddle sorter closeup

size is greatly decreased. The PS overall height is about 115 mm, including the height of rotary solenoid (51 mm) and paddle (64 mm), significantly shorter than rotary disk (300 mm). Hence, the overall height of the grading and sorting system decreases from 1220 mm to 960 mm, satisfying the requirement of shortening the overall height by 150 mm. Third, the hydraulic/mechanical power transmission system has been vastly simplified by eliminating synchronization requirements (Fig. 5.6b). In the PS design, the hydraulic motor directly powers the screw conveyors via a chain, and all other

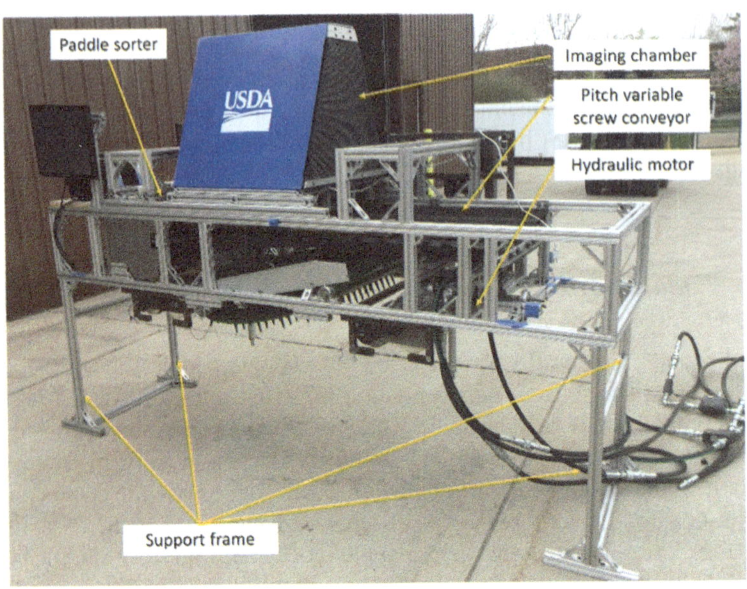

Fig. 5.6 Apple infield grading and sorting system with paddle sorter incorporated. Frame is for support and display and would be removed before it is incorporated into a harvest platform

components (e.g., angle gearbox and couplers) used in the RS are avoided. The apple infield grading and sorting system was constructed, with the newly-designed PS incorporated (Fig. 5.6).

5.2.5 *Laboratory Tests of the Apple Infield Grading and Sorting System*

5.2.5.1 Bruising Evaluation

'Red Delicious' (RD) and 'Golden Delicious' (GD) were manually picked during two harvest seasons (2017 and 2018) from an orchard (Lansing, MI, U.S.) and then carefully transported to the lab and stored in a cold storage room (35 °F) before they were used for the tests. These apples were used for all the experiments in this study. Bruising evaluations of RS and PS were conducted in 2018 and 2019, respectively.

The RS was tested at three speeds—2.0, 2.5, and 3.0 apples s^{-1}, while 2.5, 3.0, and 3.5 apples s^{-1} for the PS. Compared to the PS, the RS was tested at lower speeds due to its poor performance at 3.5 apples s^{-1} based on preliminary tests. For each speed of sorter, 40 apples per variety were used to test the system. The 40 apples were randomly divided into in four groups, with 10 apples/group. Apples were fed onto the screw conveyor and caught at the end of the sorter gently. For RS, both Gates 1 and 2 were at close mode (phantom lines in Fig. 5.3); for the PS, the paddle was closed (Fig. 5.5c) during the test. Both settings were intended to let apples collide maximal times with the system.

Apples were taken out of the cold storage room 4 h before the experiment and kept at room temperature. Each apple was visually checked carefully to identify and mark (black color marker) existed bruises before experiment. After the experiment, apples were stored at normal room temperature for 24 h, and then they were peeled and visually observed to identify newly generated bruises. Apples were graded into different levels based on the USDA fresh market apple standard (Table 5.1).

Table 5.1 Classification of apple bruise damage [37]

Class	USDA fresh market standards	Bruise specification
1	"Extra fancy"	No bruising
2	"Extra fancy"	Bruise diameter ≤ 3.2 mm (1/8 in.)
3	"Extra fancy"	Bruise diameter 3.2 to 6.4 mm (1/ 8 to 1/4 in.)
4	"Extra fancy"	Bruise diameter 6.4 mm (1/4 in.) to 12.7 mm (1/2 in.) or area of several bruises < 127 mm^2
5	"Fancy"	Bruise diameter 12.7 to 19 mm (1/2 to 3/4 in.) or total area of multiple bruises < 283 mm^2
6	Downgraded	Bruises larger than the tolerances in "Fancy"
7	Downgraded	Cuts or punctures of any size

5.2.5.2 Grading System Repeatability

The repeatability tests include intra- and inter-lane repeatability. For the intra-lane repeatability, a certain apple goes through one lane of the grading system twice to get two grading results. If the two results are the same, it is recorded as "1"; otherwise, it is recorded as "0". Finally, the apple went through all three lanes. For the inter-lane repeatability, a certain apple goes through the three lanes one by one, and then the three grading results are compared. If the three results are the same, it is recorded as "1"; otherwise, it is recorded as "0". The repeatability coefficient (RC) is calculated as the average of results. A high RC value indicates a good repeatability of the grading system; otherwise, it demonstrates its poor performance.

RD and GD (120 apples per cultivar) were used for evaluating the grading system repeatability at three speeds (2.5, 3.0, and 3.5 apples s^{-1}). The 120 apples of each cultivar were randomly separated into four groups, with each group 30 apples. They were first used for intra-lane repeatability test, and then for inter-lane test. Since each apple was used multi-times, a dedicated person stood at the end of screw conveyor to catch apples gently to avoid bruising.

5.2.5.3 Sorting Accuracy

Sorting accuracy is defined as the consistency between the grading result and the apple destination—fresh market and processing apples should arrive at their corresponding bins. For both sorters, algorithms graded apples into two grades—fresh market apples to Bin 2 and processing apples to Bin 1 (Figs. 5.3 and 5.5). The Bin 3 of RS was not used in this experiment. Since all the three lanes were exactly the same, only one lane was selected for both mode tests. The sorting accuracy experiment comprised fruit single and continuous modes. In the single mode, one apple was fed onto the screw conveyor, and after it was collected at the bin, the next apple was fed into the system. Both grading result and destination of individual apples were recorded. In the continuous mode test, all apples were numbered at the stem location to minimize its effect on grading before running the experiment. Then, ten apples were put onto the screw conveyor in the number sequence at full load (one by one without a gap between two apples). Then, the apple infield grading and sorting system was turn on, and apples were collected at bins. If an apple graded as fresh market arrived at Bin 2, or an apple graded as processing arrived at Bin 1, it was recorded as "1"; if an apple graded as fresh market arrived at Bin 1 or an apple graded as processing arrived at Bin 2, it was recorded as "0". The sorting accuracy coefficient (SAC) was the average of these results. During the experiment, when apple puncture and cut occurred, the apple was replaced by a new one with similar size and color.

A total of 640 apples (320 of RD and 320 of GD) were used to test the infield grading and sorting system accuracies with RS and PS incorporated in 2018 and 2019, respectively. Among the 320 apples, half (160 apples) were used in the single mode test and the other half in continuous mode. The 160 apples of each cultivar were randomly separated into four groups. For each group, there were 40 apples, with each group ran 2 replications.

Table 5.2 Bruising experimental results of two different sorters

System throughput	Rotary sorter				Paddle sorter			
	Red delicious		Golden delicious		Red delicious		Golden delicious	
	Grade 1*	Extra fancy#	Grade 1	Extra fancy	Grade 1	Extra fancy	Grade 1	Extra fancy
2.0 apples $^{-}$s	5%	70%	5%	95%	NA	NA	NA	NA
2.5 apples $^{-}$s	25%	100%	0%	85%	50%	100%	70%	100%
3.0 apples $^{-}$s	5%	70%	0%	85%	55%	100%	60%	100%
3.5 apples $^{-}$s	NA	NA	NA	NA	70%	100%	35%	100%
Average	12%	80%	2%	88%	58%	100%	55%	100%

Note * and # refer to Table 5.1

5.3 Results and Discussion

5.3.1 Apple Bruising

The apple bruising results of two sorters are shown in Table 5.2. The RS resulted in average Extra Fancy (EF) rate 80% for RD and 88% for GD under the speed ranges of 2.0–3.0 apples $^{-}$s. Considering the industrial requirement of > 95% apples in EF category, the RS could not meet the requirements. One potential reason for high bruising rate was that apples dropping from the end of the screw conveyors collided the dividers of the RS. The opposite moving directions of the fruit and dividers increased the collision magnitude. In addition, if the apple was not fully discharged from the Gate 1, the next paddle would press the apple to the base, which would cause apple bruising or even cut/puncture. However, all tested apples of PS were graded as EF, meeting the industrial use requirements. Since all apples were graded as EF and more than 55% of apples were graded at Grade 1 level (no bruising), it shows the PS handles apples gently. In addition, cuts and punctures were not observed during the bruising test of PS.

5.3.2 Machine Vision System Repeatability

5.3.2.1 Intra-lane Grading Repeatability

The same-lane grading RCs at three speeds range from 92 to 94% and 90% to 91% for RD and GD, respectively (Fig. 5.7). The high values of RC (>90%) demonstrate a satisfactory performance of the intra-lane repeatability. For either variety, RCs at

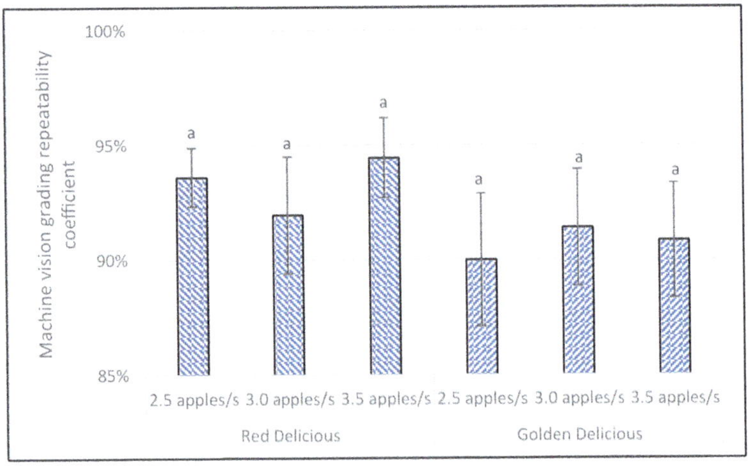

Fig. 5.7 Machine vision grading repeatability for two varieties. Error bars in the bar chart represent two standard deviations, and bars for each group with different letters are significantly different by Tukey test ($\alpha = 0.05$)

three different speeds are not significantly different from each other (same letters in each group), indicating the system speeds do not significantly affect the grading result repeatability.

Further analysis was conducted by re-grouping the data according to three speeds (Fig. 5.8). Under any of the three tested speeds, the RCs for GD and RD are not significantly different from each other, which demonstrates the machine vision system has

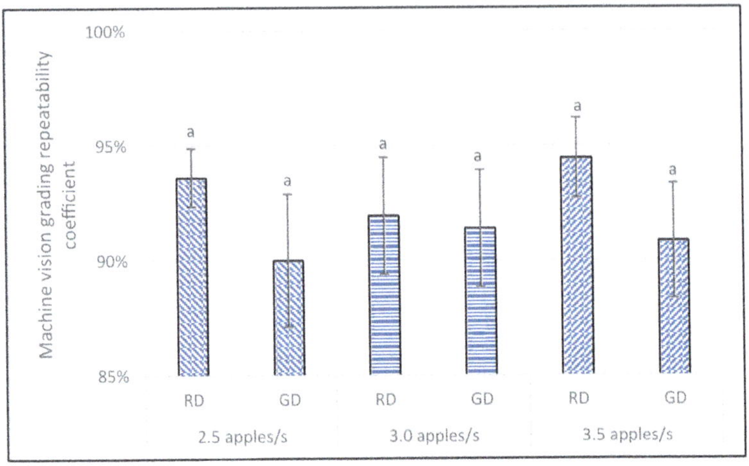

Fig. 5.8 Machine vision grading repeatability at three speeds. Error bars in the bar chart represent two standard deviations, and bars for each group with different letters are significantly different by Tukey test ($\alpha = 0.05$). RD and GD represent 'Red Delicious' and 'Golden Delicious', respectively

a robust performance for red (RD) and green color (GD) apples. The high intra-lane grading RCs demonstrate the reliability and robustness of developed algorithms in using imaging techniques for fruit size estimation.

5.3.2.2 Inter-lane Grading Repeatibility

Experimental results (Fig. 5.9) show that the RCs among three lanes range from 88 to 93% and 81% to 85% for RD and GD, respectively. The results show that for either cultivar, the system speeds would not affect the grading RC, which is the same scenario as the results of intra-lane evaluation. The high inter-lane RCs (>81%) could be attributed to the satisfactory distortion correction while developing the algorithms by our team members [24].

Further analysis was conducted by re-categorizing the results according to different speeds. For the speeds of 2.5 and 3.5 apples ¯s, the RD has a significantly higher RC over the GD, and the reason should be attributed to the existing long stems of GD apples. A majority of the long stems have a yellow color similar to the GD apples. In this scenario, the inaccurate image segmentation would lead to erroneous estimations of fruit size (Fig. 5.10).

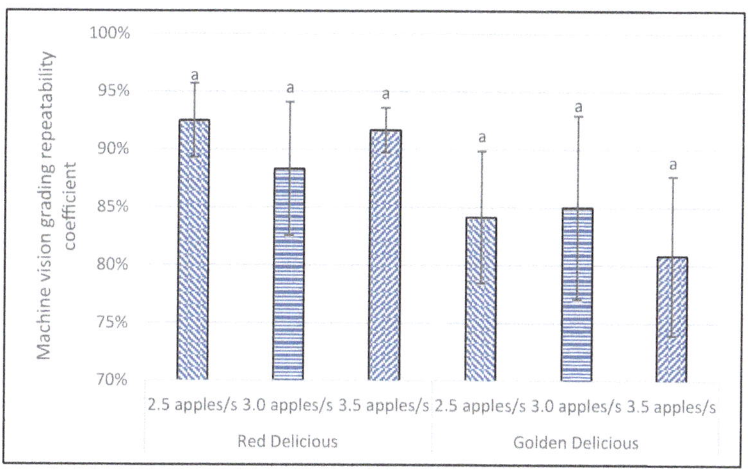

Fig. 5.9 Machine vision grading repeatability among three lanes for two varieties of apples. Error bars in the bar chart represent two standard deviations, and bars for each group with different letters are significantly different by Tukey test ($\alpha = 0.05$)

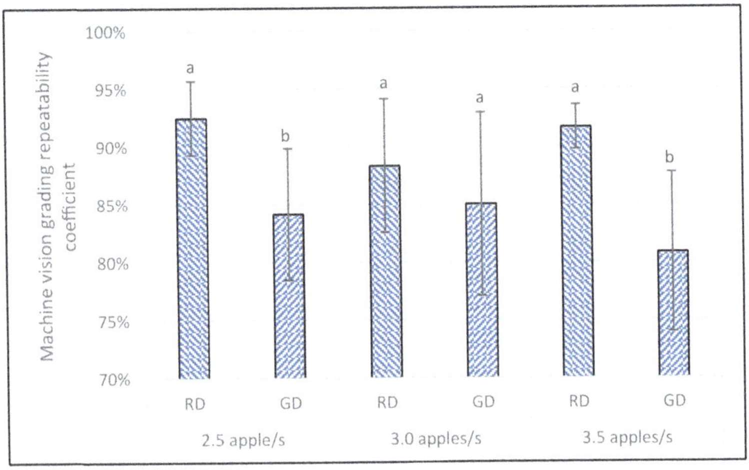

Fig. 5.10 Machine vision grading repeatability of different-lane under three speeds. Error bars in the bar chart represent two standard deviations, and bars for each group with different letters are significantly different by Tukey test ($\alpha = 0.05$). RD and GD represent 'Red Delicious' and 'Golden Delicious', respectively

5.3.3 *Sorting Accuracy*

The RS had a so poor performance, and many times in the experiment the fruit clogging issue occurred. So, the results of sorting accuracy experiment for RS would not report here. The SACs of PS for all settings (2 apple varieties × 3 throughput × 2 feeding patters) were above 99%. Thus, the sorting accuracy for the PS satisfactorily met the practical requirements.

5.3.4 *Comparison with Other Research*

This study has reported the first high throughput (10.5 apples $^{-}$s) infield use apple sorting system, and it has a number of advantages over other systems under development. First, the grading sub-system has a high throughput in image collection, processing, and decision making. The current system grades apples based on their color and size within 50 ms [24], and even with defective detection added, it is anticipated to be within 100 ms. However, some other existing systems need much more time (i.e., 400 ms) for the same process [32]. Second, the overall system is simple and compact. Using the pitch-variable screw conveyors to singulate, rotate and transport apples is a key innovation in this research. All current existing systems combine multiple mechanisms together to realize the same functions ([19, 23, 32]). Third, the paddle sorter achieved a very high sorting accuracy > 99% at the system throughput of 10.5 apples $^{-}$s, which is more satisfactory than other systems [23, 32].

5.4 Conclusions

An apple infield grading and sorting system consisting of apple singulation and rotation, grading and sorting sub-systems was developed. A rotary sorter (RS), consisting of a rotary disk and two gates, was incorporate into the infield grading and sorting system and extensive lab tests showed its unsatisfactory performance in terms of excessive bruising, bulky size, and low sorting accuracy. It was then decided to put more efforts to develop a new sorter to replace the rotary sorter. Based on the requirements of bruising free, compact design, simple mechanical and hydraulic power systems, and high sorting accuracy (>95%), a paddle sorter (PS) was developed consisting of a solenoid and a paddle. The PS has addressed the bruising issue by resulting all tested apples into Extra Fancy grade. In addition, extensive tests of the grading and sorting system with the PS incorporated showed a sorting accuracy > 99% at the speed of 10.5 apples $^{-}$s. The PS has significantly reduced the infield grading and sorting system height from 1,220 mm to 960 mm since the paddle took less space than the rotary disk. Meanwhile, by replacing the RS with PS, the hydraulic power transmission system has been simplified by eliminating two angle gearboxs and a number of sprockets, shafts, and couplers. The simplified hydraulic power transmission system increases its reliability for long-term use. Furthermore, the high values of grading repeatability coefficients for the intra-lane (>90%) and inter-lane (>81%) tests confirmed the satisfactory performance in reliability and robustness of the grading sub-system. The infield apple grading and sorting system consisting of apple singulation and rotation, grading and PS has met the requirements for infield use, and thus it has potential for commercial application.

Credit Authorship Contribution Statement Z. Zhang: Conceptualization, Investigation, Methodology, Project administration, Writing—original draft, Writing—review & editing.

Disclaimer
Mention of commercial products or orchards in this paper is only for providing factual information and does not imply endorsement of them by authors over those not mentioned.

Declaration of Competing Interest
The authors declare that they have no known competing financial interests or personal relationships that could have appeared to influence the work reported in this paper.

References

1. Zhang Z, Heinemann P, Liu J, Schupp J, Baugher T (2017c) Brush mechanism for distributing apples in a low-cost apple harvest-assist unit. Appl Eng Agric 33(2):195–201
2. Zhang Z (2015) Design, test, and improvement of a low-cost apple harvest-assist unit. Ph.D. diss. University Park: Pennsylvania State University, Dept Agric Biol Eng
3. Zhang Z, Heinemann P, Liu J, Baugher T, Schupp J (2016) Development of mechanical apple harvesting technology—a review. Trans ASABE 59(5):1165–1180

4. Zhang Z, Heinemann P, Liu J, Baugher T, Schupp J (2016b) Design and field test of a low-cost apple harvest-assist unit. Trans ASABE 59(5):1149–1156
5. Freivalds A, Park S, Lee C, Earle-Richardson G, Mason C, May JJ (2006) Effect of belt/bucket interface in apple harvesting. Int J Ind Ergon 36(11):1005–1010
6. Zhang Z, Heinemann P (2017) Economic analysis of a low-cost apple harvest-assist unit. Hort Technol 27(2):240–247
7. Mizushima A, Lu R (2011b) Development of a cost-effective machine vision system for infield sorting and grading of apples: Fruit orientation and size estimation. ASABE Paper No. 1110723. St. Joseph, Mich.: ASABE
8. Zhang Z, Pothula A, Lu R (2017b) Development and preliminary evaluation of a new bin filler for apple harvesting and infield sorting machine. Trans ASABE 60(6):1839–1849
9. Schotzko RT, Granatstein D (2005) A brief look at the Washington apple industry: Past and present. Project Report SES 04-05. Washington State University, School of Economic Sciences, Pullman, WA. Retrieved from http://www.agribusiness-mgmt.wsu.edu/agbusresearch/docs/SES04-05_BRIEF_LOOK_WAFTA.pdf
10. Zhang Z, Pothula AK, Lu R (2017a) Economic evaluation of apple harvest and in-field sorting technology. Trans ASABE 60(5):1537–1550
11. Hansen M (2009) Ready for change. Good Fruit Grower 60(17):36–37
12. Mizushima A, Lu R (2011a) Cost benefits analysis of in-field presorting for the apple industry. Appl Eng Agric 27(1):33–40
13. Regier RD, Hiebert JF (1993) Apparatus and method for sorting objects. U.S. Patent No. 5,244,100
14. Rohrbach RP, McClure WF (1978) A production capacity conveyor for small fruit sorting: the M-belt. Trans ASAE 21(6):1092–1095
15. De Vos M, Van Arkel JM (2003) Device and method for singulating from a holder with liquid and/or individualizing of fruits. U.S. Patent No. 6,655,878
16. Agrosaw (2019) Apple grading machine. Retrieved from http://agrosaw.com/apple-grading-machines/
17. Hiebert JF (1997) Method and apparatus for handling objects. U.S. Patent No. 5,626,236
18. Throop JA, Aneshansley DJ, Upchurch BL, Anger B (2001) Apple orientation on two conveyors: performance and predictability based on fruit shape characteristics. Trans ASAE 44(1):99–109
19. Throop JA, Aneshansley DJ, Anger WC, Peterson DL (2003) Conveyor design for apple orientation. ASABE Paper No. 036123. St. Joseph, Mich.: ASABE
20. Keesling TB (1965) Fruit processing method. U.S. Patent No. 3,225,892
21. Ross EE, Meissner KE (1996) Agitating apple orientor. U.S. Patent No. 5,544,731
22. Tichy OJ (1988) Apple orienting device. U.S. Patent No. 4,746,001
23. Blasco J, Aleixos N, Moltó E (2003) Machine vision system for automatic quality grading of fruit. Biosys Eng 85(4):415–423
24. Mizushima A, Lu R (2013) A low-cost color vision system for automatic estimation of apple fruit orientation and maximum equatorial diameter. Trans ASABE 56(3):813–827
25. Pothula AK, Zhang Z, Lu R (2018) Design features and bruise evaluation of an apple harvest and in-field presorting machine 61(3):1135–1144
26. Zhang Z, Pothula A, Lu R (2018) A review of bin filling technologies for apple harvest and postharvest handling. Appl Eng Agric 34(4):687–703
27. Zhang Z, Pothula A, Lu R (2019) Improvements and evaluation of an infield bin filler for apple bruising and distributions. Trans ASABE 62(2):271–280
28. Cubero S, Aleixos N, Albert F, Torregrosa A, Ortiz C, García-Navarrete O, Blasco J (2014) Optimised computer vision system for automatic pre-grading of citrus fruit in the field using a mobile platform. Precision Agric 15(1):80–94
29. Lu R, Pothula AK, Vandyke M, Mizushima A, Zhang Z (2018) System for sorting fruit. U.S. Patent 9,919,345
30. Lu R, Zhang Z, Pothula A (2017) Innovative technology for enhancing apple harvest and postharvest handling efficiency. Fruit Q 25(2):11–14

31. FruitWorld (2016) Acquisition of Compac sorting leads to major consolidation in the sorting industry. Accessed from https://fruitworldmedia.com/index.php/featured/acquisition-compac-sorting-leads-major-consolidation-sorting-industry/
32. Sofu MM, Er O, Kayacan MC, Cetişli B (2016) Design of an automatic apple sorting system using machine vision. Comput Electron Agric 127:395–405
33. Blasco J, Cubero S, Gómez-Sanchís J, Mira P, Moltó E (2009) Development of a machine for the automatic sorting of pomegranate (Punica granatum) arils based on computer vision. J Food Eng 90(1):27–34
34. Delwiche MJ, Tang S, Thompson JF (1993) A high-speed sorting system for dried prunes. Trans ASAE 36(1):195–200
35. Furniss GW (2011) Method and apparatus for sorting small food items for softness. U.S. Patent No. 7,975,853
36. Lu R, Guyer D (2010) Quality detection gets under the skin-sensor technology. Resour Mag 17(3):10–13
37. Peterson DL, Tabb AL, Baugher T, Lewis K, Glenn DM (2010) Dry bin filler for apples. Appl Eng Agric 26(4):541–549

Chapter 6
Development, Test, and Improvement of an Infield Use Bin Filler

W. Lu, Y. Ampatzidis, Zhou Zhang, and Zhao Zhang

Abstract Filling apples into a bin in the field is a bottleneck for apple production mechanization. Though a lot of bin fillers have been developed and commercially adopted, a majority are for indoor packinghouse use, which could not meet the infield requirement due to their large size, complexity, and high cost. An automatic bin filler was developed but did not result in satisfactory performance. This study focused on identifying the reasons for the poor performance of the bin filler and made improvements. The high apple bruising incidence was caused by the apple collision at the pair of foam rollers, high velocity when fruit exiting from the pinwheel, and insufficient compartments to hold one apple by one compartment. The uneven fruit distributions were mainly caused by the short pads. A second version bin filler was constructed by adding a pair of foam rollers at the top of the bin filler, adding the pinwheel compartments number from four to nine, attaching soft foams at the pads, and replacing the short pads with long ones. Experimental results showed that the second version bin filler would generate minimal apple bruising (<2%). An

W. Lu
College of Artificial Intelligence, Nanjing Agricultural University, Nanjing 210031, China
e-mail: njaurobot@njau.edu.cn

Y. Ampatzidis
Agricultural and Biological Engineering Department, Southwest Florida Research and Education Center, University of Florida, IFAS, 2685 SR 29 North, Immokalee, FL 34142, USA
e-mail: i.ampatzidis@ufl.edu

Z. Zhang
Department of Biological Systems Engineering, University of Wisconsin-Madison, Madison, WI 53706, USA
e-mail: zzhang347@wisc.edu

Z. Zhang (✉)
Department of Agricultural and Biosystems Engineering, North Dakota State University, Fargo, ND 58102, USA
e-mail: zhang.zhao.1@ndsu.edu

Key Lab of Modern Precision Agriculture System Integration Research, Ministry of Education of China, China Agricultural University, Beijing 100083, China

Key Lab of Agriculture Information Acquisition Technology, Ministry of Agriculture and Rural Affairs of China, China Agricultural University, Beijing 100083, China

Z. Zhang et al. (eds.), *Mechanical Harvest of Fresh Market Apples*, Smart Agriculture 1,
https://doi.org/10.1007/978-981-16-5316-2_6

innovative method for quantitatively analyzing apple distributions in the bin was developed, validated, and future applied in this study. The second version bin filler distributed apples more evenly in the bin over the first version bin filler. The second version bin filler, due to its satisfactory performance, has a chance for commercial infield application.

Keywords Agricultural safety · Fresh market apples · Harvest-assist · Low-cost · Occupational injuries · Seasonal workers · Ergonomic analysis

6.1 Introduction

Apple is one of the most popular fruits worldwide for its desirable taste/flavor and benefits to health [1]. Following China, U.S. is the second largest apple producer, which produces approximately 4.5 $\times$ 10^6 MT of apples in 2020 [2]. Throughout the U.S., apples are grown in 32 states by more than 5000 apple growers. With a wholesale value of apple crops about $5 billion, the estimated annual downstream value is about $15 billion. In addition to meeting the domestic consumption requirements, U.S. apples are exported to a number of countries (e.g., Mexico and Canada), with an approximate annual export value of $ 1 $\times$ 10^8. Apple production is intensively manually involved and thus costly, among which harvest and postharvest handling (i.e., storage, sorting, grading, and packing) costs account for more than 50% of the total production cost [3–5]. The continuously growing production costs gradually reduce the U.S. apple industry competitiveness domestically and internationally [6, 7]. Hence, innovative technologies aiming to reduce production costs on apple harvest and postharvest handling should be prioritized to maintain the U.S. apple industry's competitiveness.

Though a huge number of sensing and automation technologies have been successfully applied in agricultural practices [8–11], apples are still manually harvested by seasonal workers due to their susceptibility to bruising and lack of appropriate technologies [12–14]. Researchers started to develop mechanical harvest technologies decades ago, and though a lot of prototypes were developed, none of them could meet the requirements to harvest fresh market apples due to excessive bruising and high cost [11, 15–17]. In addition, apple harvest robots have been extensively studied, but due to high cost, inefficiency, and poor robustness, none of the harvest robots is or is about to be commercialized [14, 18–21]. Thus, U.S. apple growers started to take advantage of harvest platforms to increase the harvest efficiency and decrease the harvest cost [22–25]. Regarding the high postharvest handling cost, the current practice of storing all harvested apples in the same bin regardless of their quality grades is considered as the main reason [26–31]. Researchers have demonstrated that if apples could be graded, sorted, and then stored in different bins according to their quality grades, more than 50% of the postharvest handling cost could be saved.

To help apple growers achieve cost savings on harvest and postharvest handling, a research group of the U.S. Department of Agriculture, Agriculture Research Service at Lansing MI, U.S. recently developed an apple harvest and infield sorting (HIS) machine prototype [32–34]. In the HIS machine, a key part is the bin filler, which receives apples from the sorter and then uniformly distributes them into the bin [35]. Though there are a huge number of bin fillers available on the market, they cannot be incorporated into the HIS machine for two reasons. First, almost all the bin fillers are designed to be used in packing houses (indoor environment). For the indoor environment, since there is a minimal limitation on dimension and size, almost all the bin fillers are large in size and heavy [36–39]. A lot of bin fillers take advantage of water as cushion medium to avoid apple bruising, but this principle is improper for infield use. Second, there are several bin fillers have been used in field [35], but they could not work at high throughput. All the existing infield use bin fillers can work at the maximum throughput of 4–6 apples/s, but the HIS requires to work at the throughput of 9 apples/s. After reviewing all existing bin fillers and conclude that none of the existing bin fillers could be incorporated into the HIS (even with significantly modifications) machine, the group started to design and build an innovative bin filling technology for incorporation with the HIS machine. The basic requirements of the bin filler should be compact, small, high throughput (>9 apples/), and not causing apple bruising.

A key requirement during the development of the bin filler is to distribute apples uniformly in the bin. Non-uniform distribution of apples in the bin would have two major disadvantages. First, it would easily cause extra bruising. Since apples are non-uniform, due to gravity, the high-level apples in the bin would automatically roll down to the location where apple levels are low. During the rolling down process, collisions between apples would easily result in bruising. Second, during the bin filling process, a sensor would be used to monitor the apple height in the bin. If apples are uneven, the apple height information would be inaccurate, which would lead to either the early or late lifting of the bin filler. Early lifting would lead to a large distance between the bin filler and apples in the bin, and the large distance free fall for apples would generate bruising. Compared to early lifting, the late lifting would generate more disastrous results—since the bin filler cannot be lifted timely, the accumulated apples in the bin would contact apples (the gap between the apples and bin filler would no longer exist), in which condition the bin filler would malfunction.

Based on the best knowledge of authors, only a few studies have reported the design and evaluation of a bin filler [38–40]. However, none of them quantitatively evaluated the performance of bin filler on apple distribution. Manual/visual observation is used to assess the apple distributions in the bin. However, the manual approach is inaccurate and subjective. It is therefore an objective method to measure apple distributions in the bin is needed. Recent technology progress in 3D imaging has provided a new method for measuring apple distributions in the bin [41, 42]. Due to their low cost and proven satisfactory performance by a number of studies, Kinect camera (Microsoft, Redmond, Washington, U.S.) is a potential candidate to be used to quantitatively measuring apple distributions in the bin [43–46].

This paper would first present the first version bin filler and its major shortcomings identified during the tests. Then, the second version bin filler, developed as the updated of the first version, would be described, and it was incorporated into the HIS machine. It would then report the experimental results on apple bruising and distributions of both bin fillers.

6.2 Materials and Methods

6.2.1 First Version Bin Filler

The design of the first version bin filler is shown in Fig. 6.1. The first version bin filler mainly consists of guiding curtain, guiding track, foam roller, fruit guiding panel, and spinning pinwheel. After apples enter the bin filler, they will fall down by gravity, during which the guiding curtains constrain the horizontal movement freedom. After the free fall, apples arrive at the foam roller. Then the apples would decelerate and the pair of foam rollers would release apples into the guiding panel. Since the guild panel is tilted, apples would roll down automatically by gravity to the spinning pinwheel. The pinwheel would first catch apples at the bottom of guiding panel, and then distribute them in the bin uniformly. More detailed information of the first version bin filler can be referred to Zhang et al. [39]. The first version bin filler was incorporated into HIS machine and then field tested in a commercial orchard in Spartan, MI, U.S. during the 2016 harvest season. The bin filler's performance was visually observed and video recorded as well. Based on the visual observation and further video analysis, the issues associated with the bin filler were identified.

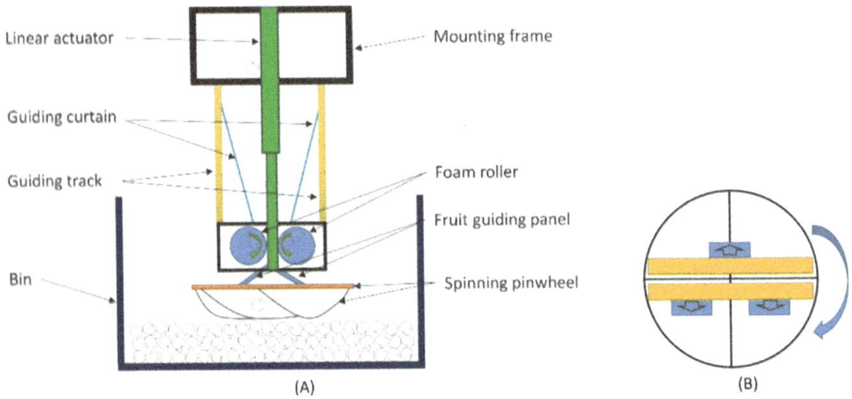

Fig. 6.1 **a** Design of first version bin filler; **b** top view of the pinwheel with four compartments

6.2.1.1 Apple Collisions at the Foam Roller

Though apples have already been singulated before entering into the bin filler, there is a large chance they would collide with each other on the foam roller, and this is because a majority of apples hit the guiding curtain, which changed apple movement trajectory. After collision with the guiding curtain, apple's movement trajectory is random and unpredictable, which resulted in apple collisions on the pair of foam rollers. To avoid this problem, it needs to find an approach to avoid apple collision with the curtain. When an apple enters the bin filler, it has a horizontal velocity, and if the horizontal velocity could be reduced to be zero, the problem could be eliminated.

6.2.1.2 Apple Collisions at the Pinwheel

The pinwheel of the first version bin filler had four pads, and it was observed that apples collided with each other on the pads. When two or more apples are in one pad, it is a challenge to avoid the collision. Thus, it is crucial to find a solution to providing each apple a compartment. There are two potential solutions to realize the concept of one apple one compartment: increase the pinwheel spinning speed, and increase the compartment number. The increase of the pinwheel spinning speed would increase apple's velocity when exiting the pinwheel, and this solution may further increase the apple incidence. Thus, increasing compartment number turns out to be a good solution.

6.2.1.3 Apple's High Velocity When Existing Pinwheel

When apples exit from the pair of foam rollers (Fig. 6.1), they only have vertical speed, and the velocity is low. The relative height between foam roller and pinwheel is about 43 cm, and while rolling down through the guiding panel, the gravitational energy is converted to kinetic energy. Though the friction between apples and guiding panel would help consume part of the energy, it was still observed that apples exiting the pads with high speed. When the high-speed apples were stopped by other apples that arrived at the bin earlier, the bruising would easily incur.

6.2.1.4 Uneven Fruit Distributions in the Bin

After completing the filling of apples into the bin, it was visually observed that the filled apples had a high 'center' but low 'corners', which means a lot of apples stayed at the center. Non-uniform fruit distribution is probably because of the design of short pads. The short pads could not carry apples far enough to the corners. A potential solution, therefore, would be to use long pad design. Though the apple pattern is visually considered as uneven, no quantitative distribution evaluations could be conducted due to the lack of appropriate tools.

6.2.2 Second Version Bin Filer

Based on the issues associated with the first version bin filler, a second version bin filler was designed and fabricated. First, a pair of foam rollers was installed at the top, immediately beneath the sorter, with a purpose to alleviate apple collisions at the bottom pair of foam rollers. The major purpose of the newly added pair of foam rollers is to release apples vertically, with minimal horizontal velocity (principle shown in the Fig. 6.2 close-up). When apples drop vertically, not hitting the guiding curtain, the issues of apple collisions at the bottom foam roller would be resolved.

Second, the sorter was operated at the throughput 9 apples s^{-1} (i.e., 3 apples s^{-1} $lane^{-1}$). Since the bin filler rotates at the speed of 1/3 revolution s^{-1} (~20 rpm), each compartment needs to handle 2.25 apples s^{-1}, indicating each compartment, at any given time, would have at least two apples. While two apples at a compartment, it would easily lead to bruising. Since it was mentioned above that increasing the spinning speed was not a good option, this study increased the compartment number from four (Fig. 6.1b) to nine (Fig. 6.2b), so that each compartment would only be responsible for one individual apple at any given time.

To address the issues of apples' high speed when discharged by the pinwheel, soft foam guides (McMaster-Carr, Aurora, Ohio) were added at the end of pads (Fig. 6.3). Apples decelerate when they collide with the guides, and thus would result in low speed when existing the pads. Preliminary results have shown that the collisions between apples and the soft foam guides would not bruise apples. Furthermore, to alleviate the issue of apples accumulated at the center of the bin and less apples distributed at the corners, the original short pads (50 cm long) were replaced with longer pads (76 cm long), so that they can carry apples longer and then release them onto the corners (Fig. 6.3).

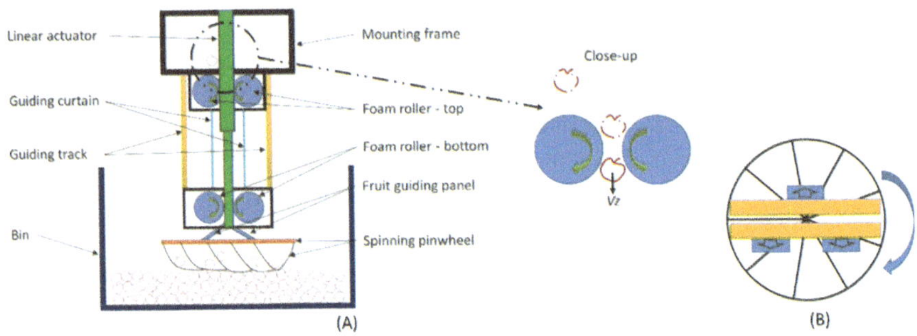

Fig. 6.2 **a** Design of second version bin filler; **b** top view of the pinwheel with nine compartments

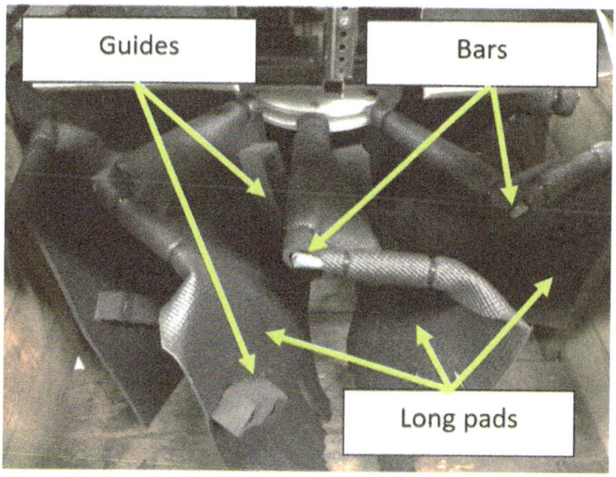

Fig. 6.3 Second version bin filler installed with nine long pads

6.2.3 *Quantification of Apples Distributions in the Bin*

Due to the lack of an appropriate method for quantifying apple distributions in the bin, an innovative approach was proposed in this study based on a depth camera (Kinect-v2, Microsoft Corp., Redmond, WA, U.S.). The camera was used to collect the depth image (512 × 424 pixels @ the rate of 30 fps). The depth camera has a field of view (FoV) of 70° × 60° (horizontal × vertical), and a normal distance measurement of 0.5 m to 4.5 m. The depth camera was mounted about 2.1 m above the bin to collect depth images.

Before using the depth camera to collect the depth images, it was first calibrated for its accuracy in depth measurement. For the calibration, 13 bars with different heights ranging from 25 to 635 mm with 51 mm increments was fabricated, with one apple placed at the top of each bar (Fig. 6.4). While placing apples at the top of the bar, their orientations were random, which can present the real conditions. Then, the height would be the bar length plus the apple height (apple height measured manually). The depth image was collected with the layout shown in Fig. 6.4. Three depth images were collected, and the average of individual pixel was used as the final depth data. The depth image represents the distance from the camera to the apples, while the manually measured height was the distance to the bottom of the bin. Then, the depth information collected by the depth camera was converted to the height to the ground by being subtracted by the distance between the bottom of the bin and the height of the depth camera to the bottom of the bin (2.1 m).

6.2.4 *Apple Bruising Evaluation*

The bin filler two was mounted onto the HIS machine and field tested in the 2017 harvest season in a commercial orchard in Sparta, MI, U.S. Workers used the platform

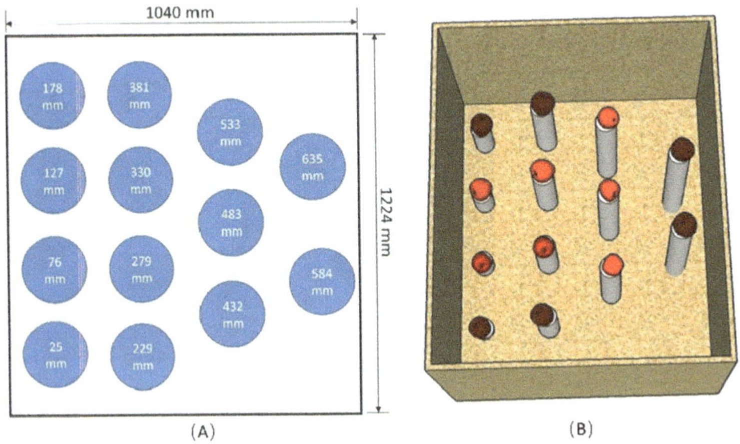

Fig. 6.4 Test setup for calibrating the depth camera's accuracy on distance measurement at different heights: **a** layout of support bars with different heights, and **b** apples placed on the top of bars with different heights

and pick apples, and all the collected apples were automatically conveyed into the bin [47, 48]. The HIS machine tested on two apple varieties: 'Gala' and 'Blondee'. The filling process started with an empty bin, and about 400 apples (samples) for each cultivar were collected from the bin for bruising evaluation. The 400 apples were randomly selected from different layers of the pile of apples, and each apple was carefully placed into cardboards trays. These trays of apples were then immediately transported to the lab in Michigan State University at Lansing, MI, U.S., where they were kept in room temperature for 24 h to allow the development of bruising. Then, each apple was manually/visually evaluated for bruising conditions using the USDA fresh market apple standards (Table 6.1).

Table 6.1 Classification of apple bruise damage [38]

Class	USDA fresh market standards	Bruise specification
1	"Extra Fancy"	No bruising
2	"Extra Fancy"	Bruise diameter $\leq$ 3.2 mm (1/8 in.)
3	"Extra Fancy"	Bruise diameter 3.2 to 6.4 mm (1/ 8 to 1/4 in.)
4	"Extra Fancy"	Bruise diameter 6.4 mm (1/4 in.) to 12.7 mm (1/2 in.) or area of several bruises < 127 mm2
5	"Fancy"	Bruise diameter 12.7 to 19 mm (1/2 to 3/4 in.) or total area of multiple bruises < 283 mm2
6	Downgraded	Bruises larger than the tolerances in "Fancy"
7	Downgraded	Cuts or punctures of any size

6.2.5 Evaluation of Apple Distribution

6.2.5.1 Experimental Setup

The sorting system has three lanes, and it was observed during the field tests that the center lane tended to transport more apples than either single side lanes [39]. A mechanism was developed to have even feeding of apples for the three lanes, and was pending for test. Thus, in evaluating the two bin fillers for their performance in fruit distributions, two patterns were tested: fruit uniform (FUF) and non-uniform (FNF) feeding.

To have a good understanding of fruit distributions, a bin of 'Gala' apples (~400 kg) were purchased from a commercial apple packinghouse in Belding, MI, U.S. After apples were taken back to the lab in Michigan State University, they were kept in cold storage at 5 °C until the test. About four hours ahead of the tests, the apples were taken out of the cold storage room, which would make them at the normal room temperature (~22 °C) during the tests.

During the tests, two persons stood on the ground to drop apples directly onto the bottom pair of foam rollers (Fig. 6.2a) at the throughput of 4 apples s^{-1}. Experiments were conducted on the two versions of bin fillers with FUF and FNF patterns. For the FUF, one worker dropped three apples at a time, with each lane one apple. For FNF, one worker dropped four apples at a time, with one, two, and one on the side, center, and side lane, respectively. There are two fruit feeding patters and two bin fillers. For each experiment, it was replicated three times, and thus, there were a total of 12 runs. During each run, 20 boxes of apples were prepared, with each box around 20 kg. Among the 20 boxes, 18 were used for the experiment, and the other 2 boxes were kept aside for replacing apples that were badly bruised or ruined during the experiment. In each run, depth images were taken three times—6, 12, and 18 boxes of apples were filled into the bin, corresponding to 1/3, 2/3, and 7/8 of the bin filled.

6.2.5.2 Imaging Processing Procedure

An image processing algorithm was developed in Matlab (MathWorks Inc., Natick, MA., U.S.) to process the acquired depth images/data. Several outlines were noticed for the collected data, and a potential reason was that there were some awkward places that cannot properly reflect the light signal back to the camera. When an outlier was found, the value of the outlier would be replaced by the average of all the neighboring pixels' values. Figure 6.5 shows the process of removing outliers.

The depth image of apples was divided into 8 × 8 blocks. Since the bin has dimensions of 1224 × 1040 mm (length × width), each block thus covers an area of 153 × 130 mm. With more apples filled in the bin, the relative distance between the apples at the top surface and the camera is reduced. Thus, each block had 28 × 24 pixels, 31 × 27 pixels, and 34 × 30 pixels, corresponding to the 1/3, 2/3, and 7/8 filling of the bin, respectively.

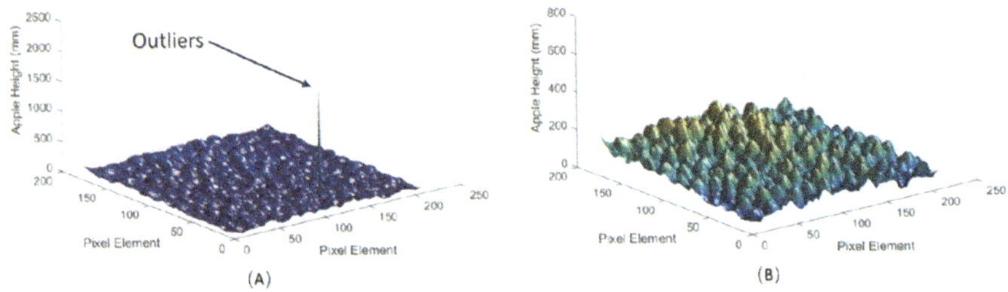

Fig. 6.5 Process of removing outliers from the original data before further process: **a** data with outliers; **b** data after removing outliers

6.2.5.3 Quantification of Apple Distributions

The height difference between blocks is also concerned. Among the 64 blocks, the block with the largest height value was considered as the base block. Then, the height difference between each block with the based block was calculated. All the 64 blocks were categorized into four groups: group 1 for the absolute difference $\leq$70 mm, group 2 for absolute difference ranging from 71 to 140 mm, group 3 for absolute difference ranging from 141 to 210 mm, and group 4 for absolute difference $\geq$ 211 mm.

6.3 Resutls and Discussion

6.3.1 Apple Bruising Evaluation Results

The infield tests of bin filler version 2 in the 2017 harvest season indicated that 99 and 98% of 'Gala' and 'Blondee' apples were categorized as Extra Fancy, which performed superior over the bin filler version 1 with only 91% 'Gala' apples categorized as Extra Fancy [39, 40]. Furthermore, the bin filler version 2 did not downgrade apples, but the bin filler version 1 downgraded about 2% of the tested apples. The less bruising caused by bin filler version 2 was mainly because of the modifications. The addition of top pair of foam rollers lowered the chance of apple collisions at the bottom pair of foam rollers. The long pads and added guide further decelerated the apples and reduced the collision magnitude, which helped lower the bruising ratio. It was also observed the nine compartment design did significantly reduce more than two apples sharing one compartment, in which way the apple-to-apple collision at one compartment is avoided.

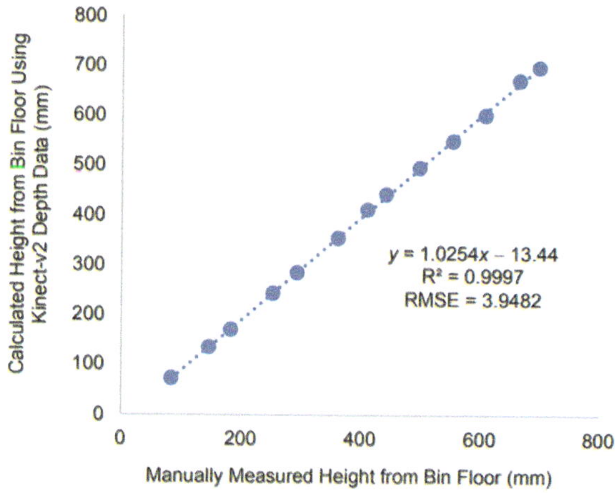

Fig. 6.6 Depth camera measurement accuracy to manually measured heights

6.3.2 *Apple Distribution Evaluation*

6.3.2.1 Depth Camera Accuracy Evaluation

The correlation between manually and depth camera measured apple heights are shown in Fig. 6.6. The difference between the two values ranged from 0.1 to 11.6 mm, with a mean absolute error of 1.3 mm and R^2 value of 0.99. Based on these results, it can be concluded that the accuracy of the depth camera can meet the requirement of being used for apple height measurement.

6.3.2.2 Overall Apple Distributions

Figure 6.7 shows the mean and standard deviation (SD) values for the fruit distributions in the bin under four different settings (e.g., 2 bin fillers × 2 feeding formats). When one third of the bin is filled, the apple distributions were neither affected by the bin filler versions (versions 1 and 2) nor by the fruit feeding patterns (FUF and FNF). A potential reason for this is that the bin has a flat bottom, and apples are fairly even because they can relatively move freely at the bottom. However, when 2/3 and 7/8 of the bin filled, bin filler version 2 led to apples more evenly distributed in the bin, which was caused by the long pad carrying apples closer to the corners. However, the feeding patterns (i.e., FUF and FNF) did not significantly affect the apple distributions.

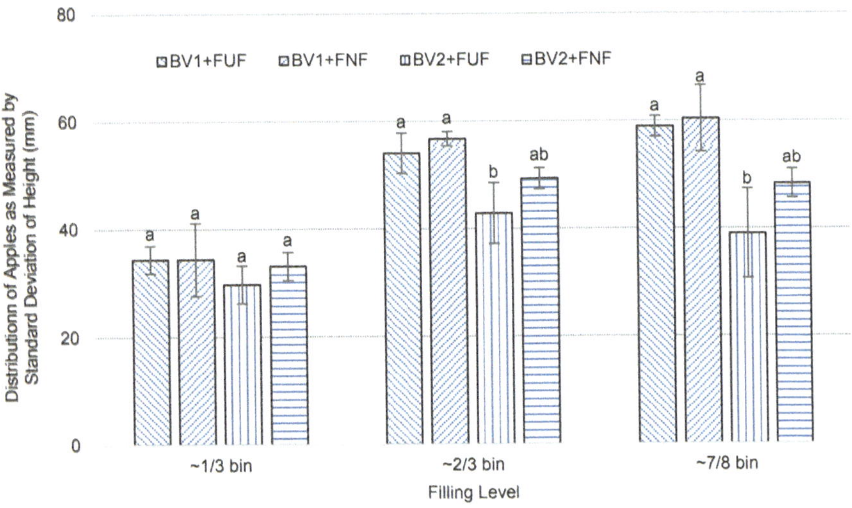

Fig. 6.7 Apple distributions in surface contour standard deviation. BV1—bin filler version 1; BV2—bin filler version 2; FUF—fruit uniform distributions; FNF—fruit non-uniform distributions; bars with different letters in each group are significantly different by Tukey's test at 0.05 significance level

6.3.2.3 2D Display of Apple Distributions

The spatial variance of fruit for 8 × 8 blocks (64 blocks) at three different filling levels is shown in Fig. 6.8. When the bin is filled with 6 boxes of apples (1/3), the maximum height difference among the 64 blocks were between 140 and 210 mm for the settings of BV1 + FUF and BV1 + FNF. However, for the settings of BV2 + FUF and BV2 + FNT, the difference was smaller, and was between 140 and 210 mm. When the bin was filled with 18 boxes apples (7/8 filled), the difference was greater than 210 mm for the settings of BV1 + FUF and BV1 + FNT, and was between 140 and 210 mm for the setting of BV2 + FUF. Thus, it can be concluded that the BV2 had better performance over BV1, demonstrating the use of long pads did improve the apple even distributions.

6.4 Conclusion

By finding the issues associated with bin filler version 1, a second version bin filler was developed by significantly modification on the first version bin filler, which includes the addition of a pair of foam rollers at the top, replacing the short pads with longer ones, attaching foam guides, and using a 9 compartments pinwheel to replace the original version with 4 compartments. Bruising evaluations and apple distributions on the two bin fillers was conducted and evaluated, and results showed that the

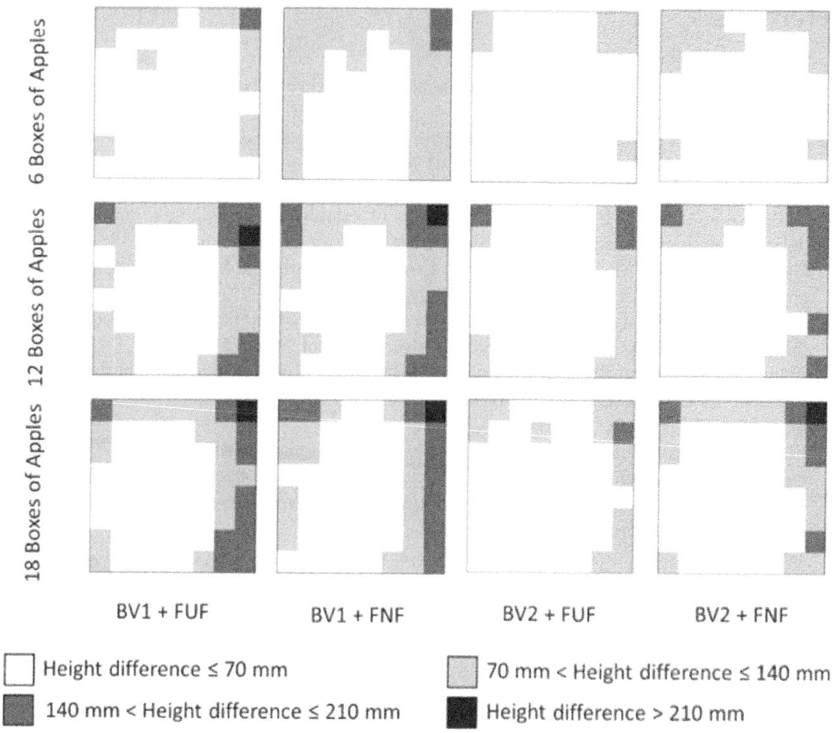

Fig. 6.8 Relative fruit height variations for 8 by 8 blocks under three different fillings and four experiment settings. Three different fillings are 6, 12, and 18 boxes of apples filled, corresponding to 1/3, 2/3, and 7/8 bin filled. BV1—bin filler version 1; BV2—bin filler version 2; FUF—fruit uniform distributions; FNF—fruit non-uniform distributions

second version bin filler had a superior performance over the first one in bruising less apples and distributing apples more evenly in the bin. It is therefore can be concluded that the modifications significantly improved the bin filler's performance. Due to the satisfactory performance of the bin filler, it has a chance to be used as a standalone system for infield apple bin filling.

CRediT Authorship Contribution Statement **W. Lu**: Writing—original draft. **Y. Ampatzidis**: Writing—review & editing. **Z. Zhang**: Writing—review & editing. **Z. Zhang**: Conceptualization, Investigation, Methodology, Project administration, Writing—original draft, Writing—review & editing.

Disclaimer
Mention of commercial products or orchards in this paper is only for providing factual information and does not imply endorsement of them by authors over those not mentioned.

Declaration of Competing Interest
The authors declare that they have no known competing financial interests or personal relationships that could have appeared to influence the work reported in this paper.

References

1. Boyer J, Liu RH (2004) Apple phytochemicals and their health benefits. Nutr J 3(1):5. https://doi.org/10.1002/9781118464663.ch11
2. US Apple Association (2021) Retrieved from https://usapple.org/
3. Galinato SP, Gallardo RK, Miles CA (2014) 2013 cost estimation of establishing a cider apple orchard in western Washington
4. Gallardo K, Galinato P (2012) Cost estimates of establishing, producing, and packing red delicious apples in Washington. Washington State University, Pullman
5. Gallardo K, Taylor M, Hinman H (2010) Cost estimates of establishing and producing gala apples in Washington. Washington State University, Pullman
6. Zhang Z (2015) Design, test, and improvement of a low-cost apple harvest-assist unit. Ph.D. diss. University Park: Pennsylvania State University, Dept Agric Biol Eng
7. Zhang Z, Heinemann PH, Liu J, Schupp JR, Baugher TA (2016a) Design and field test of a low-cost apple harvestassist unit. Trans ASABE 59(5):1149–1156. https://doi.org/10.13031/trans.59.11708
8. Flores P, Zhang Z, Igathinathane C, Jithin M, Naik D, Stenger J, Kiran R et al (2021) Distinguishing seedling volunteer corn from soybean through greenhouse color, color-infrared, and fused images using machine and deep learning. Industr Crops Prod 161:113223. https://doi.org/10.1016/j.indcrop.2020.113223
9. Jahan N, Flores P, Liu Z, Friskop A, Mathew J, Zhang Z (2020) Detecting and distinguishing wheat diseases using image processing and machine learning algorithms. ASABE Paper No. 2000372. ASABE, St. Joseph, MI. https://doi.org/10.13031/aim.202000372
10. Yao L, Hu D, Zhao C, Yang Z, Zhang Z (2021) Wireless positioning and path tracking for a mobile platform in greenhouse. Int J Agric Biol Eng 14(1):216–223. https://doi.org/10.25165/j.ijabe.20211401.5627
11. Zhang Z, Heinemann PH, Liu J, Baugher TA, Schupp JR (2016b) The development of mechanical apple harvesting technology: a review. Trans ASABE 59(5):1165–1180. https://doi.org/10.13031/trans.59.11737
12. Luo R, Lewis KM, Zhang Q, Wang SM (2012) Assessment of bruise damage by vacuumapple harvester using an impact recording device. ASABE Paper No. 121338094. ASABE, St. Joseph, MI. https://doi.org/10.13031/2013.37288
13. Zhang Z, Heinemann P, Liu J, Schupp J, Baugher T (2014) Design, fabrication, and testing of a low-cost apple harvest-assist device. ASABE Paper No. 141839738. ASABE, St. Joseph, MI. https://doi.org/10.13031/aim.20141839738
14. Zhang Z, Igathinathane C, Li J, Cen H, Lu Y, Flores P (2020) Technology progress in mechanical harvest of fresh market apples. Comput Electron Agric 175:105606. https://doi.org/10.1016/j.compag.2020.105606
15. Allshouse GW, Morrow CT (1972) Over-the-row harvester for dwarf fruit trees. Trans ASAE 15(6):1038–1043. https://doi.org/10.13031/2013.38067
16. Millier WF, Rehkugler GE, Pellerin RA, Throop JA, Bradley RB (1973) Tree fruit harvester with insertable multilevel catching system. Trans ASAE 16(5):844–850. http://dx.doi.org/https://doi.org/10.13031/2013.37641
17. Monroe GE (1982) An over-the-row continuous tree-crop harvester. Trans ASAE 25(4):888–892. https://doi.org/10.13031/2013.33634
18. Baeten J, Donne K, Boedrij S, Beckers W, Claesen E (2008) Autonomous fruit picking machine: a robotic apple harvester. In: Field and service robotics. Springer, Berlin, Heidelberg, pp 531–539. https://doi.org/10.1007/978-3-540-75404-6_51
19. Davidson JR, Silwal A, Hohimer CJ, Karkee M, Mo C, Zhang Q (2016a) Proof-of-concept of a robotic apple harvester. IEEE/RSJ Int Conf Intell Robots Syst 634–639. https://doi.org/10.1109/iros.2016.7759119
20. Davidson J, Silwal A, Karkee M, Mo C, Zhang Q (2016b) Hand-picking dynamic analysis for undersensed robotic apple harvesting. Trans ASABE 59(4):745–758. https://doi.org/10.13031/trans.59.11669

21. Tong J, Zhang Q, Karkee M, Jiang H, Zhou J (2014) Understanding the dynamics of hand picking patterns of fresh market apples. ASABE Paper No. 141898024, ASABE, St. Joseph, MI. https://doi.org/10.13031/aim.20141898024
22. Baugher T, Schupp J, Lesser K, Harsh RM, Seavert C, Lewis K, Auvil T (2009) Mobile platforms increase orchard management efficiency and profitability. Acta Hort 824:361–364. https://doi.org/10.17660/ActaHortic.2009.824.42
23. Lu R, Zhang Z, Pothula A (2017) Innovative technology for enhancing apple harvest and postharvest handling efficiency. Fruit Q 25(2):11–14
24. Schupp J, Baugher T, Winzeler E, Schupp M, Messner W (2011) Preliminary results with a vacuum-assisted harvest system for apples. Fruit Notes 76(4):1–5
25. Zhang Z, Heinemann PH (2017) Economic analysis of a low-cost apple harvest-assist unit. HortTechnology 27(2):240–247. https://doi.org/10.21273/horttech03548-16
26. Mizushima A, Lu R (2011a) Development of a cost-effective machine vision system for in-field sorting and grading of apples: Fruit orientation and size estimation. ASABE Paper No. 1110723. ASABE, St. Joseph, MI
27. Mizushima A, Lu R (2011b) Cost benefits analysis of in-field presorting for the apple industry. Appl Eng Agric 27(1):33–40. https://doi.org/10.13031/2013.36222
28. Mizushima A, Lu R (2013a) A low-cost color vision system for automatic estimation of apple fruit orientation and maximum equatorial diameter. Trans ASABE 56(3):813–827. https://doi.org/10.13031/trans.56.9343
29. Mizushima A, Lu R (2013b) An image segmentation method for apple sorting and grading using support vector machine and Otsu's method. Comput Electron Agric 94:29–37. https://doi.org/10.1016/j.compag.2013.02.009
30. Zhang Z, Pothula AK, Lu R (2016c) Economic analysis of a self-propelled apple harvest and in-field sorting machine for the apple industry. ASABE Paper No. 162456644. ASABE, St. Joseph, MI. https://doi.org/10.13031/aim.20162456644
31. Zhang Z, Pothula AK, Lu R (2017a) Economic evaluation of apple harvest and in-field sorting technology. Trans ASABE 60(5):1537. https://doi.org/10.13031/trans.12226
32. Lu R, Pothula AK, Vandyke M, Mizushima A, Zhang Z (2018) System for sorting fruit. U.S. Patent 9,919,345
33. Pothula AK, Zhang Z, Lu R (2018) Design features and bruise evaluation of an apple harvest and in-field presorting machine. Trans ASABE 61(3):1135–1144. https://doi.org/10.13031/trans.12327
34. Zhang Z, Lu Y, Lu R (2021) Development and evaluation of an apple infield grading and sorting system. Postharvest Biol Technol 180:111588. https://doi.org/10.1016/j.postharvbio.2021.111588
35. Zhang Z, Pothula AK, Lu R (2018a) A review of bin filling technologies for apple harvest and postharvest handling. Appl Eng Agric 34(4):687–703. https://doi.org/10.13031/aea.12827
36. Peterson DL, Wolford SD (2002) Mechanical harvester for tree fruits. U.S. Patent No. 6,442,920 B1
37. Peterson DL, Wolford SD (2003) Bin filler for fruit crops. U.S. Patent No. 6,644,905 B1
38. Peterson DL, Tabb AL, Baugher TA, Lewis K, Glenn DM (2010) Dry bin filler for apples. Appl Eng Agric 26(4):541–549. https://doi.org/10.13031/2013.32057
39. Zhang Z, Pothula AK, Lu R (2017b) Development and preliminary evaluation of a new bin filler for apple harvesting and in-field sorting machine. Trans ASABE 60(6):1839–1849. https://doi.org/10.13031/trans.12488
40. Zhang Z, Pothula AK, Lu R (2017c) Development of a new bin filler for apple harvesting and in-field sorting with a review of existing technologies. ASABE Paper No. 1700662. ASABE, St. Joseph, MI. https://doi.org/10.13031/aim.201700662
41. Jiang L, Koch A, Scherer SA, Zell A (2013) Multi-class fruit classification using RGB-D data for indoor robots. In: Proceedings of the IEEE international conferences on robotics and biomimetics (ROBIO). IEEE, Piscataway, NJ, pp 587–592. https://doi.org/10.1109/ROBIO.2013.6739523

42. Jiang Y, Li C, Paterson AH (2016) High-throughput phenotyping of cotton plant height using depth images under field conditions. Comput Electron Agric 130:57–68. https://doi.org/10.1016/j.compag.2016.09.017
43. Azzari G, Goulden M, Rusu R (2013) Rapid characterization of vegetation structure with a Microsoft Kinect sensor. Sensors 13(2):2384–2398. https://doi.org/10.3390/s130202384
44. Nguyen TT, Vandevoorde K, Wouters N, Kayacan E, De Baerdemaeker JG, Saeys W (2016) Detection of red and bicoloured apples on tree with an RGB-D camera. Biosyst Eng 146:33–44. https://doi.org/10.1016/j.biosystemseng.2016.01.007
45. Rosell-Polo JR, Gregorio E, Gene J, Llorens J, Torrent X, Arno J, Escola A (2017) Kinect v2 sensor-based mobile terrestrial laser scanner for agricultural outdoor applications. IEEE/ASME Trans Mechatron 22(6):2420–2427. https://doi.org/10.1109/TMECH.2017.2663436
46. Wang Z, Walsh KB, Verma B (2017) On-tree mango fruit size estimation using RGB-D images. Sensors 17(12):2738. https://doi.org/10.3390/s17122738
47. Zhang Z, Pothula AK, Lu R (2018b) Improvements and evaluation of an infield bin filler for apple bruising and distributions. ASABE Paper No. 1800921. ASABE, St. Joseph, MI. https://doi.org/10.13031/aim.201800921
48. Zhang Z, Heinemann PH, Liu J, Schupp JR, Baugher TA (2017d) Brush mechanism for distributing apples in a lowcost apple harvest-assist unit. Appl Eng Agric 33(2):195–201. https://doi.org/10.13031/aea.11971

Chapter 7
Economic Analysis of an Apple Harvest and Infield Sorting Machine

Zhaohua Zhang, Y. Ampatzidis, L. Fu, and Zhao Zhang

Abstract U.S. apple industry is facing challenges as the high production cost is hurting its competitiveness nationally and internationally. Since apple harvest and postharvest handling account for a majority (>50%) of the production cost, researchers developed innovative technologies to reduce the harvest cost and lower the fruit postharvest handling fee. An innovative apple harvest and infield sorting (HIS) machine was developed by combining harvest platform and infield sorting technologies. This study conducted the economic analysis on the HIS machine by comparing machine cost and cost savings generated by adopting the machine. The economic evaluation was conducted based on the machine-induced harvest efficiency increase of 43, 53, and 63%, and an operation of 360 h per harvest season. The machine can benefit both fresh market and processing apple growers. For fresh market apple orchards, when the processing apple incidences ranging from 5 to 15%, the annual benefits by adopting the HIS would range from $13,500 to $78,400 for the machine price between $100,000 and $160,000. With the same price range,

Z. Zhang
College of Economics and Management, Shandong Agricultural University, Tai'an 271018, China
e-mail: zzhang@sdau.edu.cn

Y. Ampatzidis
Agricultural and Biological Engineering Department, Southwest Florida Research and Education Center, University of Florida, IFAS, 2685 SR 29 North, Immokalee, FL 34142, USA
e-mail: i.ampatzidis@ufl.edu

L. Fu
College of Mechanical and Electronic Engineering, Northwest A&F University, Yangling 712100, Shaanxi, China
e-mail: fulsh@nwafu.edu.cn

Z. Zhang (✉)
Key Lab of Modern Precision Agriculture System Integration Research, Ministry of Education of China, China Agricultural University, Beijing 100083, China
e-mail: zhao.zhang.1@ndsu.edu

Key Lab of Agriculture Information Acquisition Technology, Ministry of Agriculture and Rural Affairs of China, China Agricultural University, Beijing 100083, China

Department of Agricultural and Biosystems Engineering, North Dakota, State University, Fargo, ND 58102, USA

Z. Zhang et al. (eds.), *Mechanical Harvest of Fresh Market Apples*, Smart Agriculture 1,
https://doi.org/10.1007/978-981-16-5316-2_7

benefits of adopting the HIS machine for a processing apple orchard with processing apple incidence between 80 and 90% ranges from $23,900 to $81,700. The economic analysis could help apple growers in deciding whether to purchase the HIS machine, and provide baseline information for dealers when they introduce the HIS machine to apple growers. By improving labor productivity and reducing fruit postharvest handling cost, the HIS machine brings benefits to U.S. apple growers, and thus is promising for commercialization.

Keywords Harvest platform · Infield sorting · Economic analysis · Fresh market apples

7.1 Introduction

Apple is one of the most popular fruits in the U.S. and worldwide, and can be consumed in multiple ways, such as fresh eating, apple juice, and apple sauce [1]. Currently, apples are still manually harvested, and the ladder-bucket method is commonly used [2–4]. During the harvest process, workers wear a bucket, which is used to temporarily hold apples. When the bucket is full, workers walk to a bin to release apples [5]. Though workers can stand on the ground to pick low-level apples, they need to take advantage of ladders to get access to high-level apples. Therefore, for the conventional ladder-bucket harvest method, workers have to conduct a lot of activities that are not directly related to apple picking, such as transporting apples to the bin, releasing apples, and moving ladders, leading to low efficiency. Additionally, the ladder-bucket method could easily lead to occupational injuries [6]. Due to the shrinking labor pool, increasing labor cost, and strict immigration policy, the labor-intensive ladder-bucket harvest method is facing a great challenge [7]. For these reasons, researchers intended to explore alternative harvest methods, and mechanical harvest became a focus research area. With respect to mechanical harvest of fresh market apples, many different methods have been proposed and tested, such as shake-and-catch method and air jet principle [3]. Though a number of apple mechanical harvest prototypes have been developed, none of them ends up with commercial application due to excessive bruising and high cost. Simultaneously, several apple harvest robotics have been developed and tested. However, they are still far from commercialization for the low reliability, high cost, and technical bottlenecks [8]. In response to the rising labor shortage and increased labor cost, researchers and practical engineers start to work on the harvest platform concept [9]. Working on the harvest platform, workers can spend more time on apple picking by eliminating some irrelevant activities, such as moving, climbing, and descending ladders, which in turn improves harvest productivity [10]. In additional, by avoiding the use of buckets, the strength requirements of apple harvest are significantly lowered—a fully filled bucket weights about 20 kg. The harvest platform technology can be used as an immediate alternative of the conventional ladder-bucket apple harvest method before the harvest robots can be commercially applied.

Complexity of apple postharvest handling causes a high proportion of handing cost in apple production. Researchers have reported that the cost on postharvest handling (i.e., storage, grading, sorting, and packing) accounts for about 30–40% of the total production cost [11]. Since the high production cost lowers the competition of U.S. apple industry nationally and worldwide, it is crucial to adopt innovative technologies to reduce the postharvest handling cost [12]. Currently, all harvested apples with mixed quality grades are stored into the same bins. The fully filled bins are then transported into the packinghouses, where apples are stored. When the orders from retailers arrive, apples would be pulled out of the storage and then go through the grading, sorting, and packing line, after which fresh market apples are delivered to supermarkets for sale, and processing apples are used for making juice or sauce [13]. Considering the fresh market apples can sell at a good price while the processing apples would only sell cheaply, and the postharvest handling cost is high, if there is a large portion of the mixed apples at processing level, apple growers may not break even. If apples can be sorted infield with fruit of the same quality stored in the same bin (a bin would either be all fresh market apples or processing apples), apple growers would achieve significant cost savings on the postharvest handling of apples [14]. Since all existing fruit sorting technologies are designed for indoor use and cannot be used in field due to size and cost, it is necessary to develop infield apple sorting technology [15].

Though numerous apple harvest platforms have been developed and available on the market, their adoption rate is low due to the poor economic performance [14, 16]. Apple growers need to invest $50,000 to $120,000 on a platform, which can only be used for apple harvest for 2–3 month a year. The limited economic benefits have reduced apple growers' enthusiasm on adopting the harvest platform technology. If a machine could combine the harvest platform and apple infield sorting technology, it would greatly improve the economic benefits to apple growers compared to the harvest platform alone. Based on this concept, a research team of the U.S. Department of Agriculture, Agricultural Research Service at Lansing, MI proposed the development of an apple harvest and infield sorting (HIS) machine [12]. The proposed HIS machine can work as a platform, and simultaneously it would grade and sort apples in the field according to their quality (i.e., color and size), and the store the same quality apples into the same bin. The last few years, this machine has been developed, evaluated and further improved. The satisfactory performance during multiple years' field tests showed that it could be used commercially.

Availability of technology does not guarantee apple growers' willingness to adopt the technology, and the adoption of the innovative technology mainly relies on its economic performance. Thus, this study would focus on the economic evaluation of the newly developed fresh market apple HIS machine. Specific objectives of the study were to: (1) calculate the annual machinery cost, including ownership and operating cost, (2) estimate cost savings by adoption of the HIS machine due to increased harvest efficiency and decreased postharvest handing cost, and (3) compare the machinery cost and cost savings to provide suggestions on adoption of the apple HIS machine.

7.2 Materials and Methods

7.2.1 Apple Harvest and Infield Sorting Machine

The HIS mainly consists of harvest platform, sorting system, bin filling, and automatic bin handling, which could accommodate a harvest crew of six workers. Among the six workers, two work on the ground and four on the platform. The workers only need to pick apples and then place them onto the conveyors. The conveyor would then transport apples to the machine vision-based sorting system. The machine vision system would first evaluate apples according to their quality in terms of size and color. According to customers' requirements, different thresholds can be preset before using the machine, and each apple would be categorized as either fresh market or processing. Then, the results (i.e., fresh market or processing) from the machine vision system would trigger corresponding mechanical system to guide apples to different bins, which realizes the concept of same quality apples stored in the same bin (Fig. 7.1).

To increase the automatic level of the whole system, a bin filler should be developed. The bin filler is responsible to catch apples exiting from the sorter, and then deliver them to the bin with minimal bruising. Another key requirement of the bin filler is to be able to work satisfactorily at a high throughput (>6 apple s^{-1}). Additionally, the bin filler should be able to work reliably for a long-term use. An innovative bin filler was developed, tested, improved, and finally incorporated into the HIS machine with satisfactory performance.

For all the currently existing platforms, when the bins are full, workers need to manually replace the full bins with empty ones. The replacement of bins would take

Fig. 7.1 Apple harvest and infield sorting machine under infield test in a commercial orchard

2–4 min, and during this process, workers have to idle until the replacement procedure is completed. The downtime has significantly reduced the overall harvest efficiency. To shorten the downtime, a bin handling system was developed and incorporated into the machine. While a bin is fully filled, the bin handling algorithm would be triggered—the full bin would first be removed, and then an empty bin carried by the machine would be transported to the original full bin location. The bin handling system has been developed, and field tests demonstrated its reliable performance. So far, the HIS machine has been developed and tested, and it is ready for commercial use.

7.2.2 *Annual Machine Cost Estimation*

Considering the apple harvest window is a little longer than one month in MI, U.S., the HIS machine was assumed to run 36 days per season. Assuming a 10 h/day working time, the machine operates a total of 360 h/year. The annual HIS machine cost consists of two parts: annual ownership and operating costs [17–21]. The annual ownership costs have nothing to do with the use of the machine, and the operating costs are determined by the amount of machine use.

7.2.2.1 Ownership Cost

The annual ownership cost, also named as fixed cost, can be calculated using the following Eqs. (7.1 and 7.2).

$$C_A = P_M \times C_0 \tag{7.1}$$

$$C_0 = \frac{1 - S_V}{L} + \frac{1 + S_V}{2} \times i + K_2 \tag{7.2}$$

The description of each parameter shown in Eqs. (7.1 and 7.2) is shown in Table 7.1. The ownership cost includes depreciation, interest, taxes, insurance, and housing. The housing cost represents shelter and maintenance equipment to ensure fewer repairs in the field and less deterioration (i.e., mechanical components and appearance from weathering) [19]. In this study, the HIS machine life was assumed to be 10 years, the salvage value factor was selected to be 0.1; annual interest was estimated to be 0.07. The parameter of K_2 representing tax, housing, and insurance, was estimated to be 0.02. Based on these known parameters, C_0 was calculated as 0.1485.

Table 7.1 Parameters used for calculating ownership cost

Parameter	Description
C_A	Annual ownership cost
P_M	Machine purchase price
C_O	Ownership cost coefficient
L	Machine life (years)
S_V	Machine salvage value factor at the end of the machine life
i	Annual interest rate
K_2	Ownership cost factor, including taxes, housing, and insurance

7.2.2.2 Operating Cost

To maintain the HIS machine to operate normally, owners need to invest on fuel, lubrication, repair and maintenance. The operating cost is thus closely associated with the machine operating time—the more time the HIS machine is operated, the more investment on fuel, lubrication, repair, and maintenance would occur. In this study, the operating cost was estimated based on ASABE standards EP496.3 [18], which is determined by machine use hours, repair and maintenance cost, fuel cost, and lubrication cost.

The annual repair and maintenance cost can be calculated using (Eqs. 7.3 and 7.4), and parameters are explained in Table 7.2. Based on the given information that machine life is 10 years, the total working hours for the HIS machine would be 3600 h (360 h/year × 10 years). After knowing the total working hours, based on the ASABE standards EP496.3 [18], the RF_1 and RF_2 was determined as 0.11 and 0.17, respectively. Finally, C_1 was calculated as 0.017 using (Eq. 7.4).

$$C_{rm} = P_M \times C_1 \tag{7.3}$$

$$C_1 = RF_1 \times \left(\frac{h}{1000}\right)^{RF_2} \tag{7.4}$$

Table 7.2 Parameter used to calculate annual repair and maintenance cost

Parameters	Description
C_{rm}	Accumulated annual repair and maintenance cost
C_I	Accumulated annual repair and maintenance cost coefficient
RF_1	Repair and maintenance factor 1
RF_2	Repair and maintenance factor 2
h	accumulated annual use of machine

The self-propelled apple HIS machine is powered by a 28 hp gasoline engine, and the fuel consumption cost could be calculated based on Eq. (7.5). The Q_{avg} represents the average gasoline consumption cost, and the P_{pto} stands for maximum power take-off (PTO). The gasoline price was estimated to be \$0.58/L, and the PTO was assumed to be the same as the engine power (28 hp). Based on these information, the Q_{avg} was calculated as \$3.64/h. Given the assumption that lubrication (L_{avg}) cost was 15% of the fuel cost (Eq. 7.6; [19]), the lubrication cost was calculated as \$0.55/h.

$$Q_{avg} = 0.13 \times P_{pto} \tag{7.5}$$

$$L_{avg} = 0.15 \times Q_{avg} \tag{7.6}$$

7.2.2.3 Total Annual Machine Cost

The total annual machine cost includes annual ownership cost and the operating cost, and according to the above calculations, both are related to the machine price. Though machine price is a key factor for the total annual machine cost, the actual price for this machine is not yet determined. After reviewing a bunch of commercialized harvest platforms, this study provided a reasonable price for the HIS machine. The Munckhof harvest platform (Munckhof, Horst, The Netherlands) sells at the price of \$78,400 and the vacuum apple harvester (Phil Brown Welding Corp., Conklin, MI, U.S.) can be purchased at \$126,500 [22]. Considering both platforms do not have infield sorting system incorporated, the HIS machine price was set ranging from \$60,000 to \$160,000.

7.2.3 *Annual Cost Savings on Harvest Labor*

7.2.3.1 Harvest Productivity Increase by Decreased Occupational Injuries

Apple harvest is one of the nation's most hazardous agricultural practices [23]. During the apple harvest process, workers are exposed to a number of musculoskeletal disorders, ladder falls, and dermatitis [24–26]. Earle-Richardson et al. [27, 28] reported that the morbidity of occupational injuries for orchard workers was about 4%, among which the leading injuries were back, neck, and shoulder strains/sprains, followed by the ladder falls. Demers and Rosenstock [29], Husting et al. [30], McCurdy et al. [31], and Brower et al. [32] reported that the strains/sprains accounted for about 1/3 of all injuries, ranging from 31 to 39%. In this study, the strains/sprains morbidity was estimated as 35%. Picking apples on the platform, workers do not need to bend or stretch their bodies, and could work in comfortable postures [21], and thus it

Table 7.3 Parameters used to calculate productivity reduction caused by strains/sprains

Parameter	Description	Coefficient
PR_{ss}	Productivity reduction caused by strains/sprains	
M_{oi}	Morbidity of occupational injuries (4%)	0.04
M_{ss}	Morbidity of strains/sprains (35%)	0.35
P_{rss}	Productivity reduction caused by strains/sprains (42%)	0.42

can be reasonably assumed that strains/sprains were eliminated. The strains/sprains could reduce the harvesting productivity by 42% [11, 33]. Then, the total productivity reduction by sprains/strains was obtained as 0.59% (Eq. 7.7). Description of each parameter is summarized in Table 7.3.

$$PR_{ss} = M_{oi} \times M_{ss} \times P_{rss} \tag{7.7}$$

Following sprains/strains, ladder falls are the second most serious injury type for orchard workers and can cause serious consequences, such as fractures [34–36]. Research on apple harvest employees at Washington state orchards reported that the top three ladder-related injuries were strains/sprains (38%), contusions (26%), and fractures/dislocations (12%) [37]. Ladder fall accidents usually lead to employees off for the full harvest season [38, 39]. By working on the HIS machine, the use of ladders is eliminated, as well as the ladder-related injuries. In this study, it is assumed that when ladder fall occurs to a worker, the worker would miss the whole harvest season (36 days). When contusions occur to workers, they are assumed to result in the same harvest productivity loss/reduction to workers as the sprains/strains. Thus, the harvest productivity reduction by ladder falls accidents was calculated as 0.19% based on Eq. (7.8). Explanations for each parameter are descried in Table 7.4.

$$PR_{lf} = M_{oi} \times M_{lf} \times \left(M_{lss} \times PR_{lss} + M_{lrc} \times PR_{lrs} + M_{lfd} \times PR_{lfd}\right) \tag{7.8}$$

Table 7.4 Parameters used to calculate productivity reduction caused by ladder falls

Parameter	Description	Coefficient
PR_{lf}	Productivity reduction caused by ladder falls	
M_{oi}	Morbidity of occupational injuries	0.04
M_{lf}	Morbidity of ladder falls in the occupational injuries	0.12
M_{lss}	Morbidity of strains/sprains caused by ladder falls	0.38
PR_{lss}	Productivity reduction caused by strains/sprains from ladder falls	0.42
M_{lrc}	Morbidity of contusions caused by ladder falls	0.26
PR_{lrc}	Productivity reduction caused by contusions from ladder falls	0.42
M_{lfd}	Morbidity of fracture/dislocation caused by ladder falls	0.12
PR_{lfd}	Productivity reduction caused by fracture/dislocation from ladder falls	1

By adopting the HIS machine to reduce the occupational injuries, workers' harvest productivity could be increased by 0.9%. The productivity increase is calculated based on Eq. (7.9), where HP_i represents the harvest productivity increase. Due to the limited data resources, other serious consequences induced by ladder falls, such as death, were not incorporated into the calculation. Thus, in this study, it was assumed that harvest productivity would increase 3% (about three times of the calculated HP_i) by using the platform to replace ladders.

$$Hp_i = \frac{1}{1 - PR_{ss} - PR_{lf}} - 1 \tag{7.9}$$

7.2.3.2 Harvest Productivity Increase by Using the Platform

Since the developed HIS machine has not gone through extensive field tests, its performance on harvest productivity increase is unknown. Therefore, apple harvest productivity increase by using the HIS machine was evaluated by reviewing existing commercial harvest platforms. Researchers and practical engineers have reported that different harvest platforms could improve harvest productivity ranging from 15 to 60% [16, 40, 41]. After comparing the detailed HIS machine with those platforms on the market, a conservative estimate has been reached—the HIS machine would improve harvest efficiency ranging from 40 to 60%.

7.2.3.3 Harvest Labor Cost

Washington is the major fresh market apple production state in the U.S., and thus, it would be a major state for the application of HIS machine. In Washington state, apple harvest employees are paid by piece rate—\$66 MT^{-1}, which was used as harvest labor cost in this study.

7.2.3.4 Total Labor Cost Savings

In the economic evaluation of HIS machine, workers are assumed to be paid by working hour. By adopting the HIS machine, the harvest cost would go down because workers can harvest more apples in an hour. The above detailed analysis has shown that the reduced occupational injuries increase harvest productivity by 3%, and adopting the HIS machine would increase harvest productivity by 40 to 60%. Then, three harvest productivity increase of 43, 53 and 63% (corresponding to 40, 50, and 60% increase by the platform, plus 3% increase by reduced occupational injuries) were further investigated. Since the harvest labor cost using ladder-bucket method was estimated to be \$3,089/ha for a yield of 124 bins/ha (\$66 MT^{-1}), the

Table 7.5 Labor cost savings under different harvest productivity increase

Harvest productivity increase (%)	Labor cost savings ($/ha)	Labor cost savings ($/MT)[#]	Labor cost savings ($/bin)
43	930	20	8
53	1070	23	9
63	1200	26	10

[#] Data based on apple production of 124 bin/ha (50 bin/acre) with each bin containing 0.378 MT of apples

Table 7.6 Harvest rate and machine capacity under different harvest productivity increase

	Harvest productivity increase 43%	Harvest productivity increase 53%	Harvest productivity increase 63%
Harvest rate (6 workers)	6.0	6.4	6.8
Machine capacity (MT/season)	1177.2	1259.5	1341.8

labor cost would be reduced to $2160, $2019, and $1895 per ha if harvest productivity increase by 43%, 53%, and 63%, respectively (Table 7.5).

7.2.3.5 Machine Capacity

A professional apple harvest employee can pick one bin of apples in an hour using the ladder-bucket method. Assuming seven apples weighs 1 kg, workers can pick at the rate of 0.7 apples/s. The machine sorting capacity for a harvest crew of six workers is shown in Table 7.6. Since the sorting system was designed to work maximally at the throughput of 9 apples/s, its capacity can be then calculated as 1764 MT/season. With the maximum machine capacity of 1341 MT/season (harvest productivity increase of 63%; Table 7.6), the sorting system could meet the practical requirement.

Apple growers in the U.S. can be categorized into two groups: fresh market apple growers and processing apple growers. For fresh market apple growers, the sort-out rate of the sorting system is the number of sorted-out processing apples divided by the total number of processing apples. For processing apple growers, the sort-out rate is the number of sorted-out fresh market apples by the total number of fresh apples.

7.2.3.6 Annual Cost Savings for Postharvest Handling of Procesing Apples

Based on the literature review and personal communication with two commercial packinghouses (Riveridge Packing LLC and Elite Apple Co. LLC), the costs of different postharvest handling services used in this study are listed in Table 7.7.

Table 7.7 Costs of different packinghouse services

Packinghouse service	Cost ($/MT)	Cost ($/bin)
Controlled atmosphere storage (long-term storage)	80	30
Cold storage (short-term storage)	30	12
Grading/sorting/packaging	290	110

Processing apples are assumed to be kept in cold storage for short-term (maximally one or two months), and they would not go through the sorting/grading/packing lines. However, processing apples are stored in the controlled atmosphere storage, under which condition apples can be kept for up to a year. When orders come, the apples would be taken out from the controlled atmosphere, and then they would go through the sorting/grading/packaging line.

The sorting system was able to achieve a 90% sort-out rate for processing apples, which means 10% of processing apples would still be in the mixed quality apples (fresh market + processing). For the mixed quality apples, the postharvest handling cost would be \$370/MT, including controlled atmosphere storage of \$80/MT and grading/ sorting/packaging of \$290/MT. When 10 MT apples with 40% processing apples delivered to the packinghouse without pre-sorting, the postharvest handling cost would be \$3700 (\$370/MT × 10 MT). After adopting the infield sorting technology, 3.6 MT (10 MT × 40% × 0.9) processing apples would be sorted out infield. For these 3.6 MT sorted-out processing apples, the postharvest handling fee would be \$108 (3.6 MT × \$30/MT). The remaining 6.4 MT apples of mixed quality would cost \$2,368 (6.4 MT × \$370/MT). Then, the total cost would be \$2476 (\$108 + \$2368). By adopting the infield sorting technology, the cost savings would be \$1224 (\$3700–\$2476).

For fresh market apple growers, the harvested apples would have more fresh market apples over processing apples. Table 7.8 shows the economic benefits by adopting the infield sorting technology under different processing apple ratios and yields.

Table 7.8 Apple postharvest handling cost savings ($) by using infield sorting technology

Yield (MT/ha)	Processing apple ratio 10%	Processing apple ratio 20%	Processing apple ratio 30%	Processing apple ratio 40%
20	612	1224	1836	2448
25	765	1530	2295	3060
30	918	1836	2754	3672
35	1071	2142	3213	4284
40	1224	2448	3672	4896
45	1377	2754	4131	5508

7.3 Resutls and Discussion

The infield sorting technology would benefit both fresh market and processing apple growers by increasing harvest efficiency. For fresh market apple growers, they would use the sorting technology to sort out processing apples, so that they can save cost on postharvest handling. For processing apple growers, infield sorting technology would sort out fresh apples, which would otherwise be sold as processing apples.

7.3.1 *Annual Machine Cost*

The annual machine cost is relevant to the machine price. With the HIS machine price goes up from $60,000 to $160,000, the annual machine cost ranges from $11,497 to $28,096. The ownership and repair/maintenance costs increase linearly with the machine price goes up, while the costs of fuel and lubrication remain unchanged, as they are only related to the machine operating hours. Table 7.9 summaries the annual machine cost under different machine price. The ownership cost accounts for more than 80% of the total cost.

Table 7.9 Annual machine cost under different machine price ($)

Machine price	Ownership cost	Repair and maintenance cost	Fuel cost	Lubrication cost	Total cost
60,000	8,910	1,049	1,337	201	11,497
70,000	10,395	1,224	1,337	201	13,157
80,000	11,880	1,399	1,337	201	14,817
90,000	13,365	1,574	1,337	201	16,477
100,000	14,850	1,749	1,337	201	18,137
110,000	16,325	1,924	1,337	201	19,787
120,000	17,820	2,099	1,337	201	21,457
130,000	19,305	2,273	1,337	201	23,116
140,000	20,790	2,448	1,337	201	24,776
150,000	22,275	2,623	1,337	201	26,436
160,000	23,760	2,798	1,337	201	28,096

7.3.2 Benefits for Fresh Market Apple Growers

Orchards producing fresh market apples are more intensively managed, which means they have relatively low rates of processing apples. To evaluate the economic benefits for fresh market apple growers in a practical manner, the processing apple incidence (PAI) is considered as maximally to be 20%. The sorting out rate is 90%. The economic analysis results by adopting the HIS machine under three different productivity increase ratios (i.e., 43, 53, and 63%) are shown in Fig. 7.2. For a given harvest productive increase, the higher PAI, the more benefits it would occur by adopting the HIS machine. When the PAI is 0, the sorting system does not work, and it would generate the lowest economic benefits. For a given machine price and PAI, the machine would bring more benefits with higher harvest productivity growth due to increased cost savings on labor. Assuming that machine price ranging from \$60,000 to \$160,000, PAI ranging from 0 to 20%, and harvest productivity increase ranging from 43 to 63%, the calculated benefits by adopting HIS machine have a range from −\$4600 to \$105,500. The lowest value occurs when the machine price is \$160,000, PAI is 0, and harvest productivity increase rate is 43%, and he highest value occurs when the machine price is \$60,000, PAI is 20%, and harvest productivity increase rate is 63%.

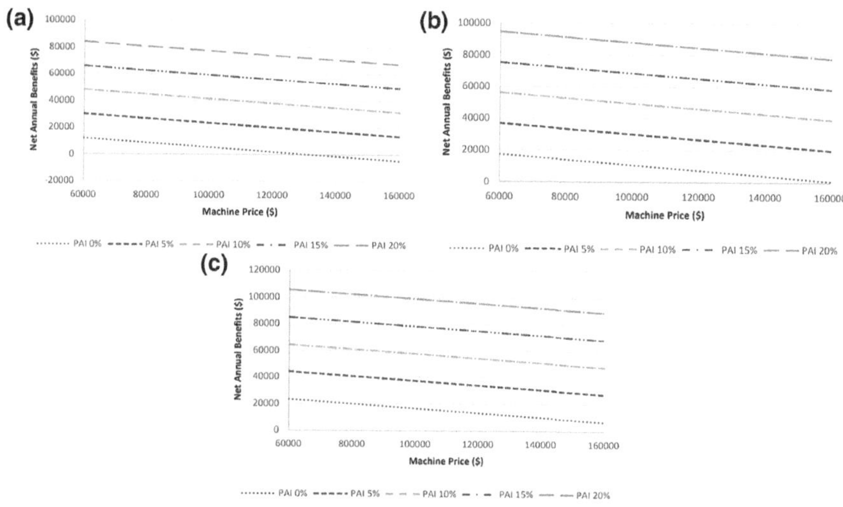

Fig. 7.2 Net annual benefits for fresh market apple growers by adopting the harvest and infield sorting machine with different processing apple incidences (PAIs): **a**, **b**, and **c** represents 43%, 53%, and 63% harvest productivity increase, respectively

7.3.3 Benefits for Processing Apple Growers

Processing apple growers sell all apples as processing apples, and those good quality apples, which meet the requirements of fresh market, sell as processing apples as well. However, by adopting the sorting system, the high quality apples could be sorted out. Thus, the price difference between fresh market apples and processing apples would be the benefits to apple growers by using the sorting system. The sorting system could sort out 90% of the fresh market apples. Since the fresh market and processing apples sell at \$0.81 kg^{-1} and \$0.21 kg^{-1}, respectively [42, 43], apple growers can earn an extra \$0.60 kg^{-1} when fresh market apples are sorted out. Since a majority of the apples are processing apples, the PAIs are determined ranging from 70 to 100%. When the PAI is 100%, it means there are no fresh market apples, and the sorting system does not work. The economic analysis results using the HIS machine for processing apples are shown in Fig. 7.3. When the machine price ranges from \$60,000 to \$160,000 and the PAI ranges from 70 to 100%, the net annual benefit ranges from −\$4,600 to \$120,000. For a given harvest productivity increase and PAI, the benefits decrease with the increase of the machine price. For a given harvest productivity increase and machine price, the benefits would decrease with the increase of PAI (less fresh market apples available for sorting).

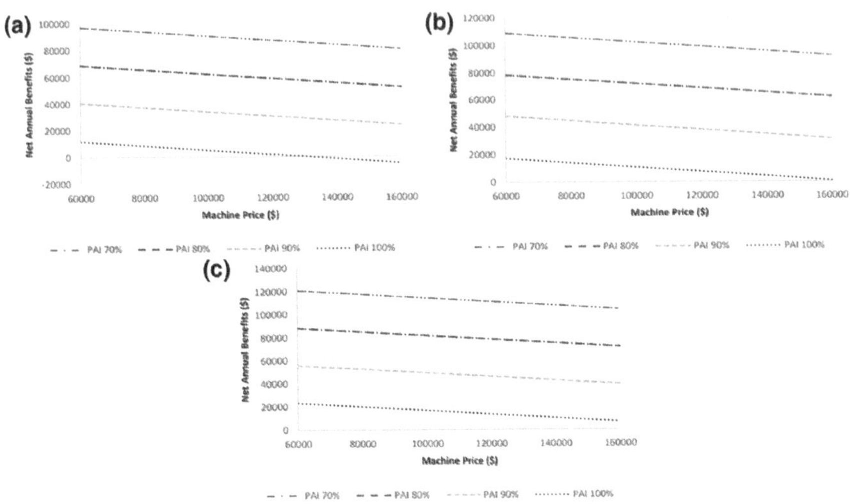

Fig. 7.3 Net annual benefits for processing apple growers by adopting the harvest and infield sorting machine with different processing apple incidences (PAIs): **a**, **b**, and **c** represents 43%, 53%, and 63% harvest productivity increase, respectively

7.4 Conclusion

Economic analysis for a newly developed apple harvest and infield sorting machine (HIS) was conducted. Among the four cost components of ownership, repair/maintenance, fuel, and lubrication, ownership cost accounts for more than 80% of the total cost. Since the ownership cost is mainly determined by the HIS machine price, it is important to set a reasonable price for the machine—too high price would reduce the willingness of apple growers to purchase the HIS machine; too low price would lower profits of the manufacturing company and dealer. The sorting system would benefit either types of apple growers—for fresh market apple growers, they would use the sorting system to sort out processing level apple, and save their cost on postharvest handling; for processing apple growers, they would use the sorting system to sort out fresh market apples, which would otherwise be sold as processing apples. Detailed analysis on processing apple incidence (PA), machine price, and harvest efficiency improvement was conducted, and the analysis results would be used either by apples growers and HIS machine dealer. For the apple growers, they can refer to the analysis results to understand the economic benefits of adopting the technology; for the HIS machine dealer, they would use the analysis results as the baseline data when they introduce the HIS machine to apple growers. In general, by combing harvest and infield sorting technology, the HIS machine would increase revenue of apple growers.

In this study, only apple harvest and in-field sorting functions of the HIS machine are considered. However, the machine could be also involved with other orchard activities, such as tree pruning, thinning, and training. In addition, the machine could also be tested to harvest other fruits, such as pears and oranges. With more functions used, the economic benefits induced by using the HIS machine would increase.

Credit Authorship Contribution Statement **Z. Zhang**: Writing—original draft. **Y. Ampatzidis**: Writing—review & editing. **L. Fu**: Writing—review & editing. **Z. Zhang**: Conceptualization, Investigation, Methodology, Project administration, Writing—original draft, Writing—review & editing.

Disclaimer
Mention of commercial products or orchards in this paper is only for providing factual information and does not imply endorsement of them by authors over those not mentioned.

Declaration of Competing Interest
The authors declare that they have no known competing financial interests or personal relationships that could have appeared to influence the work reported in this paper.

References

1. PBH (2015) State of the plate: 2015 study on America's consumption of fruit and vegetables. Produce for Better Health Foundation, Hockessin, DE. Retrieved from http://www.pbhfoundation.org/pdfs/about/res/pbh_res/State_of_the_Plate_2015_WEB_Bookmarked.pdf

2. Freivalds A, Park S, Lee C, Earle-Richardson G, Mason C, May JJ (2006) Effect of belt/bucket interface in apple harvesting. Intl J Ind Ergonomics 36(11):1005–1010. https://doi.org/10.1016/j.ergon.2006.08.005
3. Zhang Z, Heinemann PH, Liu J, Baugher TA, Schupp JR (2016a) The development of mechanical apple harvesting technology: a review. Trans ASABE 59(5):1165–1180. https://doi.org/10.13031/trans.59.11737
4. Zhang Z, Heinemann PH, Liu J, Schupp JR, Baugher TA (2016b) Design and field test of a low-cost apple harvest assist unit. Trans ASABE 59(5):1149–1156. https://doi.org/10.13031/trans.59.11708
5. Zhang Z, Heinemann PH, Liu J, Schupp JR, Baugher TA (2017b) Brush mechanism for distributing apples in a lowcost apple harvest-assist unit. Appl Eng Agric 33(2):195–201. https://doi.org/10.13031/aea.11971
6. Zhang Z, Pothula A, Lu R (2017c) Development of a new bin filler for apple harvesting and infield sorting with a review of existing technologies. ASABE Paper No. 1700662. ASABE, St. Joseph, MI. https://doi.org/10.13031/aim.201700662
7. Zhang Z (2015) Design, test, and improvement of a low-cost apple harvest-assist unit. Ph.D. diss. Pennsylvania State University, Department of Agricultural and Biological Engineering, State College, PA. Retrieved from https://etda.libraries.psu.edu/catalog/26275
8. Zhang Z, Igathinathane C, Li J, Cen H, Lu Y, Flores P (2020) Technology progress in mechanical harvest of fresh market apples. Comput Electron Agric 175:105606. https://doi.org/10.1016/j.compag.2020.105606
9. Schupp J, Baugher T, Winzeler E, Schupp M, Messner W (2011) Preliminary results with a vacuum assisted harvest system for apples. Fruit Notes 76(4):1–5
10. Zhang Z, Heinemann P, Liu J, Schupp J, Baugher T (2014) Design, fabrication, and testing of a low-cost apple harvest-assist device. ASABE Paper No. 141839738. ASABE, St. Joseph, MI. https://doi.org/10.13031/aim.20141839738
11. Zhang Z, Pothula AK, Lu R (2017a) Economic evaluation of apple harvest and in-field sorting technology. Trans ASABE 60(5):1537–1550 https://doi.org/10.13031/trans.12226
12. Pothula AK, Zhang Z, Lu R (2018) Design features and bruise evaluation of an apple harvest and in-field presorting machine. Trans ASABE 61(3):1135–1144. https://doi.org/10.13031/trans.12327
13. Zhang Z, Pothula A, Lu R (2016c) Economic analysis of a self-propelled apple harvest and infield sorting machine for the apple industry. ASABE Paper No. 2456644. ASABE, St. Joseph, MI. https://doi.org/10.13031/aim.20162456644
14. Zhang Z, Lu Y, Lu R (2021) Development and evaluation of an apple infield grading and sorting system. Postharvest Biol Technol 18:111588. https://doi.org/10.1016/j.postharvbio.2021.111588
15. Lu R, Pothula AK, Mizushima A, VanDyke M, Zhang Z (2018) U.S. Patent No. 9,919,345. U.S. Patent and Trademark Office, Washington, DC
16. Lu R, Zhang Z, Pothula A (2017) Innovative technology for enhancing apple harvest and postharvest handling efficiency. Fruit Q 25(2):11–14
17. ASABE (2011a) D497.7: Agricultural machinery management data. ASABE, St. Joseph, MI
18. ASABE (2011b) EP496.3: Agricultural machinery management. ASABE, St. Joseph, MI
19. Edwards W (2015) Estimating farm machinery costs. Iowa State University Extension, Ames IA. Retrieved from https://www.extension.iastate.edu/agdm/crops/html/a3-29.html
20. Edwards W, Ayres G, Boehlje M (2005) Estimating farm machinery costs. Iowa State University Extension, Ames IA. Retrieved from http://www.aae.wisc.edu/aae320/Budgets/IA%20Farm%20Mac hinery%20Cost%20PM710.pdf
21. Zhang Z, Heinemann PH (2017) Economic analysis of a low-cost apple harvest-assist unit. HortTechnology 27(2):240–247. https://doi.org/10.21273/horttech03548-16
22. Jones R (2015) The state of mechanical apple harvesting. Growing Produce (23 Feb. 2015). Retrieved from http://www.growingproduce.com/fruits/the-state-of-mechanicalapple-harvesting/

23. Leigh JP, McCurdy SA, Schenker MB (2001) Costs of occupational injuries in agriculture. Public Health Rep 116(3):235–248
24. Coye MJ (1985) The health effects of agricultural production: I. The health of agricultural workers. J Public Health Policy 6(3):349–370. https://doi.org/10.2307/3342402
25. Litchfield MH (1999) Agricultural work-related injury and ill health and the economic cost. Environ Sci Pollut Res 6(3):175–182. https://doi.org/10.1007/bf02987623
26. Purschwitz MA, Field WE (1990) Scope and magnitude of injuries in the agricultural workplace. Am J Ind Med 18(2):179–192. https://doi.org/10.1002/ajim.4700180210
27. Earle-Richardson GB, Brower MA, Jones AM, May JJ, Jenkins PL (2008) Estimating the occupational morbidity for migrant and seasonal farmworkers in New York State: a comparison of two methods. Ann Epidemiol 18(1):1–7. https://doi.org/10.1016/j.annepidem.2007.07.092
28. Earle-Richardson G, Jenkins PL, Freivalds A, Burdick P, Park S, Lee C, May JJ et al (2006b) Laboratory evaluation of belt usage with apple buckets. Am J Ind Med 49(1):23–29. https://doi.org/10.1002/ajim.20239
29. Demers P, Rosenstock L (1991) Occupational injuries and illnesses among Washington State agricultural workers. Am J Publ Health 81(12):1656–1658. https://doi.org/10.2105/AJPH.81.12.1656
30. Husting EL, Geiser CR, Summerill KF, Cervantes Y, Moltrum R, Ruiz C, Osorio AM (1997) Occupational agricultural injury surveillance in California: Preliminary results from the nurses using rural sentinel events (NURSE) project. J Agromed 4(3–4):269–283. https://doi.org/10.1300/J096v04n03_11
31. McCurdy SA, Samuels SJ, Carroll DJ, Beaumont JJ, Morrin LA (2003) Agricultural injury in California migrant Hispanic farm workers. Am J Ind Med 44(3):225–235. https://doi.org/10.1002/ajim.10272
32. Brower MA, Earle-Richardson GB, May JJ, Jenkins PL (2009) Occupational injury and treatment patterns of migrant and seasonal farmworkers. J Agromed 14(2):172–178. https://doi.org/10.1080/10599240902799715
33. Taylor DK, Green N, Physio D (2008) What are the productivity losses caused by musculoskeletal disorders (MSDs)? A review of the current literature. Wellnomics Ltd, Christchurch, New Zealand. Retrieved from http://www.workpace.com/assets/Uploads/WhitePapers/Wellnomics-white-paper-What-are-the-productivitylosses-caused-by-musculoskeletal-disorders.pdf
34. Lombardi DA, Smith GS, Courtney TK, Brennan MJ, Kim JY, Perry MJ (2011) Work-related falls from ladders: A follow-back study of U.S. emergency department cases. Scandinavian J Work Environ Health 37(6):525–532. https://doi.org/10.5271/sjweh.3174
35. Smith GS, Timmons RA, Lombardi DA, Mamidi DK, Matz S, Courtney TK, Perry MJ (2006) Work-related ladder fall fractures: Identification and diagnosis validation using narrative text. Accid Anal Prev 38(5):973–980. https://doi.org/10.1016/j.aap.2006.04.008
36. Stevens JA, Corso PS, Finkelstein EA, Miller TR (2006) The costs of fatal and non-fatal falls among older adults. Injury Prev 12(5):290–295. https://doi.org/10.1136/ip.2005.011015
37. Hofmann J, Snyder K, Keifer M (2006) A descriptive study of workers' compensation claims in Washington State orchards. Occup Med 56(4):251–257. https://doi.org/10.1093/occmed/kql018
38. Axelsson P-O, Carter NE (1995) Measures to prevent portable ladder accidents in the construction industry. Ergonomics 38(2):250–259. https://doi.org/10.1080/00140139508925102
39. Cohen HH, Lin L-J (1991) A scenario analysis of ladder fall accidents. J Saf Res 22(1):31–39. https://doi.org/10.1016/0022-4375(91)90011-J
40. Robinson T, Sazo MM (2013) Recent advances of mechanization for the tall spindle orchard system in New York State: Part 2. New York Fruit Q. 21(3):3–7
41. Robinson T, Hoying S, Sazo MM, DeMarree A, Dominguez L (2013) A vision for apple orchard systems of the future. New York Fruit Q 21(3):11–16
42. USDA (2015) Agricultural prices. Washington, DC: USDANASS. Retrieved from http://usda.mannlib.cornell.edu/usda/nass/AgriPric/2010s/2015/AgriPric-12-30-2015.pdf
43. Wheat, D (2014) Large apple crop holds down processing prices. Capital Press (6 Sept. 2014). Retrieved from http://www.capitalpress.com/Orchards/20140906/large-applecrop-holds-down-processing-prices

Chapter 8
Comparison and Evaluation of Apple Harvesting Process Under Different Harvest Methods

Zhaohua Zhang and Zhao Zhang

Abstract Due to decreased availability of seasonal labor and increased labor cost for apple harvest, it is necessary to explore a new harvest method to replace the conventional ladder-and-bucket approach. As ladder use is one of the main reasons which cause low efficiency of the ladder-and-bucket approach, harvest platforms have been commercialized and adopted by pioneer orchard growers to avoid the use of ladder. However, the adoption process is slow because apple growers are doubtable about the efficiency improvement brought by harvest platforms. It is, therefore, necessary to compare harvest efficiencies between the ladder-and-bucket method and harvest platforms, as well as finding potential directions for further efficiency improvement. This study first recorded videos of workers' harvesting process under different methods—ladder-and-bucket, DBR, and Huron, and then divided and categorized the continuous harvesting process into different activities for analysis. Experimental results demonstrated picking time percentages (PTPs) of 64%, 78%, and 83% for the ladder-and-bucket, DBR, and Huron, respectively. The picking activity was further evaluated in terms of approaching, detaching, and transporting apples. The results showed that detachment time percentage (DTP) was 32%, 30%, and 31%, respectively, for the three methods. Overall efficiency (OE) and overall time index (OTI) were introduced to further compare the three methods more holistically. The OEs for the three methods were 21%, 23%, and 26%, respectively, and OTIs were 45%, 71%, and 80%, respectively. Using OE as a reference, the DBR and Huron achieved an increase of 10% and 24% over the ladder-and-bucket, respectively, and an increase

Z. Zhang
College of Economics and Management, Shandong Agricultural University, Tai'an 271018, China
e-mail: zzhang@sdau.edu.cn

Z. Zhang (✉)
Key Lab of Modern Precision Agriculture System Integration Research, Ministry of Education of China, China Agricultural University, Beijing 100083, China
e-mail: zhang.zhao.1@ndsu.edu

Key Lab of Agriculture Information Acquisition Technology, Ministry of Agriculture and Rural Affairs of China, China Agricultural University, Beijing 100083, China

Department of Agricultural and Biosystems Engineering, North Dakota, State University, Fargo, ND 58102, USA

Z. Zhang et al. (eds.), *Mechanical Harvest of Fresh Market Apples*, Smart Agriculture 1,
https://doi.org/10.1007/978-981-16-5316-2_8

of 58 and 78% when using OTI as a reference. Compared to the high values of PTPs (64–83%) with limited scope for improvement, the low values of detachment time percentage (1/3 of picking time) for all three methods indicated a clear potential for improvement. Future research efforts should be directed to the investigation and improvement of the overall harvesting process, apple picking component activities (detachment and transport), reducing hand activities, and automated harvesting methods through innovative design and development.

Keywords Apple · Ladder-and-bucket · Platform · Vacuum · Time and motion study · Efficiency improvement

8.1 Introduction

Apple is one of the most popular fruits around the world consumed in fresh and processed forms. Even though the consumption of fresh fruits is recommended for a healthy diet, the field harvesting process of apples encounters several challenges, including labor shortage, physical strength demand, mechanization handling, and automation. The current and common apple harvest method relies on ladders and buckets [1–3]. As the apple tree is tall, workers can get access to the fruits based on their reach while standing on the ground, and they need to use the ladders to harvest apples in the tree at different height levels [4, 5]. In addition, workers have to wear a bucket during the entire harvesting process for the purpose of temporarily holding apples [6]. The ladder-and-bucket method is low in efficiency because workers have to conduct a number of activities that are not related to picking, such as moving/climbing/descending ladders [7, 8]. Besides this method is highly demanding of physical effort because workers have to wear a bucket (~3 kg itself and ~ 20 kg when full) during the entire harvesting process, they have to move the ladder or climb/descend the ladder with the bucket (empty to filled) frequently. The effort requirement worsens the labor shortage as only part of workers in the labor pool are suitable for this job. Furthermore, the ladder-and-bucket apple harvest method is prone to occupational injuries, such as ladder falls and musculoskeletal disorders [9–11]. Strains and sprains are also common occupational injuries during apple harvest [12]. When workers are intended to pick apples in a distance while standing on a ladder, they probably undergo some awkward postures (e.g., twisting, stretching, and bending bodies), which increase the incidence of occupational injuries [13–16]. To overcome the listed shortcomings of the ladder-and-bucket harvest approach, new harvest methods need to be developed [4, 17–19].

Researchers and engineers have been exploring new ideas to improve or even replace the ladder-and-bucket method over the past decades. Efforts have been devoted to developing apple harvest robotics to fully replace human labors [3, 20, 21]. Baeten et al. [22] and Stuntz [23] constructed apple harvest robotics with a vacuum-suction concept. Zhao et al. [24] developed an apple harvest robot with a spoon-shaped end-effector (apple gripping mechanism simulating fingers of hand while picking) for apple detachment, with a demonstrated harvest efficiency of 4 apples/min and a 77% successful picking rate. Davidson et al. [20, 21] and Silwal et al. [25] developed a harvest robot and validated its capacity of harvesting 10 apples/min with an 84% successful picking rate. Kahani and Dror [26] developed and patented an apple harvest robotics, but the performance was not reported [27]. All of the harvest robots face a common challenge that they are unable to pick apples grown in cluster. Since all these developed robots are bulky, complex, and slow for practical application, it is impractical and unrealistic to adopt them in apple harvest in field. Due to their large size, complex structure and slow practical application, these developed robots are impractical for in-field apple harvest.

Beyond the radical approach of using robots to entirely substitute human labors, development of harvest platforms to increase workers' harvest efficiency is another option ([4, 28–31]). By replacing the use of ladders, adoption of harvest platforms could avoid issues associated with ladder application (e.g., low efficiency and ladder fall accidents) [5]. Workers still need to wear buckets to hold apples temporarily while working with some platforms; while some other platforms use conveyors or vacuum tubes for fruit transportation to free workers from buckets [19]. A fundamental and rigorous study that investigates the apple harvesting process (manual and mechanical harvest systems) is essential to gain an understanding of the various actions that lead to efficiency improvement and design of better systems.

Time and motion study (TMS), defined as a scientific analysis method to measure time allocations of workers in conducting different activities, has been used extensively in measuring and improving workers' efficiency and productivity [32–34]. Hendrich et al. [35] applied the TMS to analyze the work process of nurses in hospitals and identified main areas for efficiency improvement. Lan et al. [36] analyzed workers' producing process on a lamp assembly line using TMS, and the improved working process increased workers' efficiency by 15%. Duran et al. [37] studied worker' efficiency using TMS and indicated there was 53% potential for productivity improvement. Moktadir et al. [38] identified workers' time-wasting activities using TMS in a production line of bags, and by adopting an improved working procedure with redundant activities eliminated, the productivity was improved by 13%. Given the versatile nature and application in several industries, TMS is a proper tool for efficiency analysis of the apple harvesting process as well.

To address the concerns of apple growers regarding the comparison of efficiencies of ladder-and-bucket and harvest platform methods, and to find scope for further harvest efficiency improvement, it is important to analyze workers' harvesting process with different apple harvesting methods to gain baseline information. Therefore, the objectives of this study are to: (1) video record workers' harvesting process under three different methods (ladder-and-bucket, DBR, and Huron); (2) divide and categorize the continuous harvesting process into different activities; (3) compare the three harvest methods' efficiencies using various efficiency parameters; and (4) determine the potential for efficiency improvement.

8.2 Materials and Methods

8.2.1 Introduction to Three Apple Harvest Methods

In this study, three apple harvest methods were selected for the evaluation of workers' harvesting process, which included ladder-and-bucket, Huron platform, and DBR (initials of the three joint inventors) vacuum harvester. The ladder-and-bucket approach is the most popular and conventional apple harvest method, while the modern methods of Huron and DBR harvest platforms are gradually adopted by pioneer apple growers recently in Europe and the US.

8.2.1.1 Ladder-And-Bucket Method

The ladder-and-bucket method has been the common and dominant approach for decades. Workers wear buckets to temporarily hold apples after they are picked from trees (Fig. 8.1a). When the bucket is full, workers must walk to a bin to unload the apples (Fig. 8.1c). For fruits at higher levels, workers cannot reach them by standing on the ground and ladders are needed. There are three major issues associated with this harvest method—low-efficiency, high strength requirement, and occupational injuries. The low-efficiency is attributed to a lot of activities that workers perform are not directly related to apple picking. One example is that workers have to climb and descend ladders to reach for the fruits at higher levels. The high physical effort requirement is due to the lifting of objects such as relocation of ladders and carrying of buckets. Workers need to lift a ladder (weighing ~ 10 kg) frequently from one location to another. The occupational injuries also compound from awkward postures of the workers, such as climbing a ladder with a full bucket and twisting the body to reach for the fruits in the zone of range. As apple harvesting is a seasonal agricultural

Fig. 8.1 Apple harvest with ladder-and-bucket method: **a** picking on the ground for low level apples, **b** picking on the ladder for high level apples, and **c** unloading full bucket of apples into a bin

activity, this can attract only seasonal workers that were already in demand for other competitive jobs and this situation contributes to further labor shortage.

8.2.1.2 Huron Harvest Platform

By replacing the use of ladders, the Huron harvest platforms provide a better working environment for workers to conduct harvest activities. Harvest platforms also eliminate the frequent relocation of ladders, which is one of the shortcomings associated with the use of ladders. Huron harvest platform could accommodate a harvest crew of eight workers—four perform picking stationed on the platform and the other four on the ground, and five bins onboard (Fig. 8.2a). For the four workers on the platform, it is convenient to unload the apples when the bucket is full into the bins behind them, but for the four working on the ground they have to walk to the platform to dump the apples into the bins. After all the five bins full, they are unloaded from the rear of the of platform to the ground and then a bin hauler loads five empty bins on the harvest platform at one time.

8.2.1.3 DBR Vacuum Harvester

DBR vacuum harvester is another option for apple growers. The DBR could accommodate five workers (four on the platform and one on the ground). Compared to the ladder-and-bucket and Huron methods that are involved with the use of buckets, the four workers on the DRB platform are free from buckets, and they only need to pick and then place apples into the vacuum tube. The long vacuum tube then transports apples into a decelerator, after which they are discharged to the bin. The single worker picking apple on the ground operates similarly to the four ground workers with Huron, and only responsible for the lower level apples (Fig. 8.3b). The ground picker walks to the bin carried by the platform, when the bucket is full, to unload the apples (Fig. 8.3c).

Fig. 8.2 Huron harvest platform: **a** four workers on the platform and other four on the ground (one worker on the ground are not in the view); **b** workers pick apple at low level while on the ground, and **c** workers unload apples into a bin onboard the platform

Fig. 8.3 DBR vacuum harvester: **a** four workers on the platform, **b** one worker on the ground, and **c** the worker on the ground dumps apples into the bin carried by the platform

8.2.2 Field data Collection and Basic Orchard Information

Field data collections were carried out in Schwallier Country Basket (Sparta, MI, US), Rennhack Orchards Market (Hart, MI, US), and Country Mill Farms (Charlotte, MI, US) for the ladder-and-bucket, Huron, and DBR during the harvest season of 2018. A camcorder (Sony Handy Cam FDR-AX53, Sony Corp., Japan) was used to record the picking process of individual workers at 120 frames per second (fps) for each harvest method. Researchers followed workers closely and recorded their harvesting process, while keeping a certain distance to avoid interference. Overall, each picker was video recorded for 10 to 15 min. For the ladder-and-bucket method, four workers were video recorded, and all of them picked both on ground and ladder (in turns); for the Huron, four workers (half of the harvest crew) were video recorded (2 on platform and 2 on ground); and for the DBR, all five pickers were recorded (four on platform and one on ground).

The test of DBR was conducted in an "old-style" orchard, with large canopy trees ($\approx$5.2 m in height), as well as wide tree in-row width ($\approx$1.5 m). The orchard pattern for the ladder-and-bucket method is new, with a smaller canopy ($\approx$3 m height) and tree in-row width ($\approx$1.2 m). The Huron system was tested in a high-density orchard, with trees structured in trellis-walls, small tree row width ($\approx$3 m) and tree in-row width ($\approx$0.9 m) (Table 8.1).

Each worker's 10–15 min video was divided into three sub-videos (three replications) each of approximately 3–5 min. Then, frames were extracted from videos at 120 fps, and the evaluation of workers' activities was conducted visually.

The general harvesting process analysis was conducted by dividing and categorizing the continuous harvesting process in the video into individual activities, along with the time percentage for each activity. The activities were then re-grouped into picking and non-picking for further analysis. Also, a detailed picking activity analysis was performed as this was the most significant harvest operation. Finally, the

Table 8.1 Basic orchard information for the tests of three different harvest methods

Tree information	Harvest method		
	Ladder-and-bucket	Huron	DBR
Cultivar	'Gala'	'Gala'	'Rome'
Apple tree height	≈ 3.0 m (10 ft)	≈ 3 m (10 ft)	≈ 5.2 m (17 ft)
Tree row width	≈ 4.3 m (14 ft)	≈ 3 m (10 ft)	≈ 4.3 m (14 ft)
Tree in-row width	≈ 1.2 m (4 ft)	≈ 0.9 m (3 ft)	≈ 1.5 m (5 ft)

Table 8.2 Categorization of workers' harvest activities under different methods

Activities	Picking or Non-picking	Ladder-and-bucket	Huron—on platform	Huron—on ground	DBR—on platform	DBR—on ground
Moving ladder	Non-picking	✓	×	×	×	×
Descending ladder	Non-picking	✓	×	×	×	×
Climbing ladder	Non-picking	✓	×	×	×	×
Walking to bin	Non-picking	✓	✓	✓	×	✓
Dumping apples	Non-picking	✓	✓	✓	×	✓
Walking away from bin	Non-picking	✓	✓	✓	×	✓
Picking on ground	Picking	✓	×	✓	×	✓
Picking on ladder	Picking	✓	×	×	×	×
Picking on platform	Picking	×	✓	×	✓	×
Moving/driving platform	Non-picking	×	×	×	✓	×

overall efficiencies of the three harvest methods were compared. All the processes involved in the analysis are subsequently presented in appropriate sections with detail. Categorizations of the harvest activities for three methods are summarized in Table 8.2.

8.2.3 General Harvesting Process Analysis

For the ladder-and-bucket, since all workers pick apples both on the ground and ladder, they conduct the same activities. However, for the Huron and DBR methods, workers on the platform perform different activities to workers on the ground (Table 8.2). Workers on the ground of Huron and DBR both wear buckets but they do not need to use ladders because they are not responsible for apples at high levels. However, workers picking on the Huron platform wear a bucket to temporarily hold apples, and when the bucket is full, they turn it around to unload apples into the bin. As apples are transported into the bin via vacuum tube and bin filler, workers on the DBR platform are free from wearing a bucket, as well as the activities related to transporting fruit into the bin (unloading apples). Since the Huron moves slowly at a constant speed automatically, there is no need for a dedicated driver; while for DBR, it is stop-and-go, and a worker spends part of the time driving the machine (Table 8.2). Additionally, workers have to move hydraulic-powered adjustable platforms in vertical/horizontal directions to get access to apples while working on the DBR platform.

8.2.4 Picking and Non-picking Analysis

Picking and non-picking are subdivided activities of the harvesting process (Table 8.2). The picking activities including picking on ground, ladder, and platform, while the non-picking consists of a lot of activities, such as climbing/descending/moving ladder, walking back and forth to the platform, unloading fruits, and driving the platform. The picking activity is desirable as it directly determines harvest efficiency; while the non-picking activities, even though essential components of the particular system, can be considered undesirable from the modern system point of view that could eliminate some of them through innovative technologies. This study would compare the picking and non-picking activity time percentages under different harvest methods.

8.2.4.1 Detailed Analysis of Picking Activity Components

The detailed analysis of picking activity components focuses on further evaluation of the picking activity in terms of approaching, detaching, and transporting apples. Approaching is defined as after picker's hand leaving the bucket or vacuum tube of DBR until touching fruit on the tree. Detaching is defined as plucking fruit from the limb, from the hands contacting an apple till the apple physically separated from the limb. Transporting is defined as the movement of the fruit after it is independent of the tree until they are put into the bucket or vacuum tube. Among the three activities, detaching is the most critical and delicate activity in the apple harvest, and a higher time percentage of detaching is desirable. Time percentage, instead of real time, was applied in this study to avoid other non-consequential factors, such as video length.

Fig. 8.4 Different scenarios in the detailed picking activity evaluation: **a** both hands have the same activity of picking, **b** one hand is picking while the other hand is transporting apples, **c** only one hand is visible and the other one is blocked, and **d** both hands are invisible (this occurs rarely)

8.2.4.2 Frame Evaluation

The picking and non-picking activities are easy to distinguish from an individual frame, as a worker usually conducts one activity at a given time. However, for the detailed analysis of picking activities, the scenario is more complex. All workers video recorded in this study used both hands for apple picking. However, their two hands may not perform the same activity. Some workers may use one hand for apple picking, while the other hand used for apple transportation (Fig. 8.4b). While observing the videos, it also happens that only one hand is visible in the view, and the other may be blocked by the worker's body, or tree trunk/branches/limbs/leaves (Fig. 8.4c). The worst condition is that both hands are blocked, making it difficult to know the actual activities performed by the hands (Fig. 8.4d). Such situations of not observing both hands cannot be avoided as the recording was in no way interfered with the worker's natural rhythm of apple picking activities. In this study, if only one arm is visible, the arm's activity is used in the analysis; if both arms visible, the more visible arm is chosen (e.g., right arm in Fig. 8.4b); if both arms invisible (this scenario occurs rarely), the researcher uses the best estimation.

8.3 Results and Discussion

8.3.1 General Harvesting Process Analysis

Time percentages of different activities for three harvest methods are summarized in Table 8.3. For the ladder-and-bucket, pickers spent 47% and 18% of total time picking on the ladder and ground, respectively. Among the non-picking activities, unloading apples took the most of the time (9% of the total time), and this was probably due to the slow nature of the activity to avoid bruising and occupational injuries. Moving ladder consumed the second most time of the non-picking activities (7% of the total time), and this was attributed to the weight of the ladder (~10 kg). Walking to the

Table 8.3 Time percentages of different harvest activities for three harvest methods

Activity	Ladder-and-bucket*	Huron—on platform**	Huron—on ground#	DBR—on platform##	DBR—on ground*#
Moving ladder	7%	N/A	N/A	N/A	N/A
Descending ladder	5%	N/A	N/A	N/A	N/A
Climbing ladder	5%	N/A	N/A	N/A	N/A
Walking to bin	5%	1%	10%	N/A	6%
Dumping apples	9%	3%	8%	N/A	7%
Walking away from bin	5%	2%	9%	N/A	8%
Picking on ground	18%	N/A	73%	N/A	79%
Picking on ladder	47%	N/A	N/A	N/A	N/A
Picking on platform	N/A	94%	N/A	78%	N/A
Moving/driving platform	N/A	N/A	N/A	22%	N/A

Note *, **, #, ##, and *# represent data is based on the average of four, two, two, four, and one picker(s), respectively. Values presented are the average of three replications for each picker

bin requires almost the same time percentage (5% of the total time) as the walking away from the bin, and this was because the workers went through the same route, even though walking to bin means carrying the heavier loaded bucket while walking away means moving with the lighter empty bucket. The same scenario occurred with climbing and descending the ladder.

With the Huron platform, workers spent a very high time percentage (94%) in picking, because they are free from the ladder use. Also, they spent relatively more time in unloading apples (3%) than walking to (1%) and away from (2%) the bin. In addition to they were very close to the bin, the other reason was that workers unload apples gently and slowly to avoid bruising. For the workers picking on the ground with the Huron, the picking time percentage (PTP) went down from 94 to 73%, mainly because of the large distance between the pickers and the bins carried by the platform. It was also observed in the video that ground workers picked apples quicker than the movement of the Huron platform, resulting in a relatively large distance between the picker and the bin.

While standing and picking on the DBR platform, the PTP was 78%, lower than that on the Huron platform, and this was because workers spent time in moving the hydraulic-powered platform in both directions (horizontal and vertical), as well as driving the machine. For the DBR, the one person picking on the ground spent 79% time in picking.

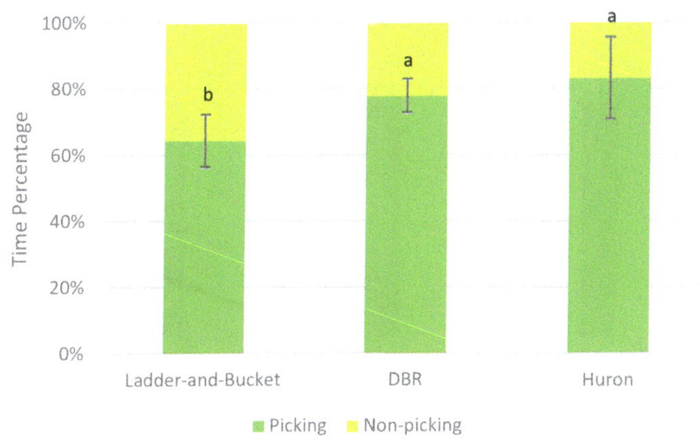

Fig. 8.5 Comparisons of picking and non-picking time percentages under different harvest methods. Error bars in the bar chart represent two standard deviations, and bars for each group with different letters are significantly different by Tukey test ($\alpha = 0.05$)

8.3.2 Picking and Non-picking Time Percentages

A comparison of the selected harvest methods' picking and non-picking activities analysis of the general harvesting process is presented in Fig. 8.5. The PTP accounts for 64, 78, and 83% of the total harvest time for ladder-and-bucket, DBR, and Huron, respectively. The results of lower PTP value of ladder-and-bucket represents a significant difference ($\alpha = 0.05$) when compared to the rest of the modern harvest methods. For the ladder-and-bucket method, the PTP of 64% was similar to the results reported by other researchers. Among the three methods, the ladder-and-bucket generated the lowest time percentage, mainly due to the handling of ladders and harvested apples. Overall, the harvest platform significantly improves the PTP from 64% (ladder-and-bucket) to 81% (average of the DBR and Huron), which accounts to a 27% improvement.

8.3.3 Detailed Analysis of Picking Activity Components

Among the components of picking activity consisting of approaching, detaching, and transporting apples, detaching was established as the essential activity. The experimental results showed the detaching time percentage is 32% (±10%), 30% (±4%), and 31% (±3%) for the ladder-and-bucket, DBR, and Huron, respectively, which were not significantly different ($\alpha = 0.05$) from each other (Fig. 8.6). Similar time percentage scenarios occurred to approaching and transporting apple activities. For all the three activities, the ladder-and-bucket method has the largest standard deviation consistently, and this is due to the fact that the trees in the orchard are in an

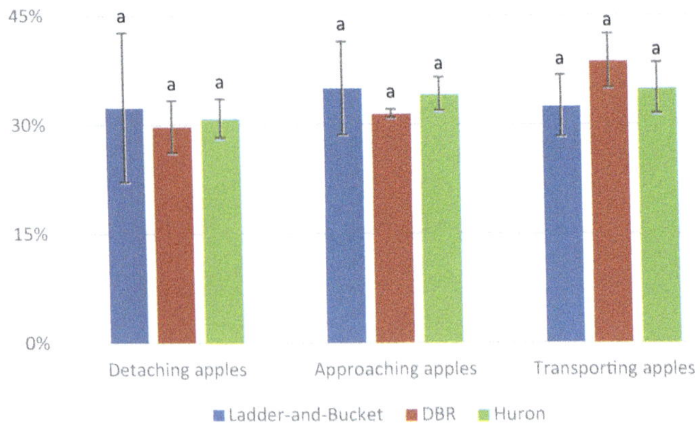

Fig. 8.6 Comparisons of time percentages for detaching, approaching, and transporting apples under different harvest methods. Error bars in the bar chart represent two standard deviations, and bars for each group with different letters are significantly different by Turkey test ($\alpha = 0.05$)

old fashion, with large tree canopies. From the picking activity detail analysis, the three methods have exhibited similar patterns in time allocations for the three sub-activities: ~ 1/3 for detaching apples, ~ 1/3 for approaching apples, and ~ 1/3 for transporting apples. This is understandable as the components of picking activity were common and independent of the harvest methods.

8.3.4 *Overall Efficiency and Overall Time Index*

Based on the picking and non-picking time analysis and the detailed analysis of picking activity components (approaching, detaching, and transporting apples), the overall efficiency (OE) is defined as the product of PTP and detachment time percentage (DTP; Eq. 8.1). From Eq. (8.1) it can be observed that increased values PTP and DTP will improve the OE. Also, while comparing the picking and non-picking time percentages, a new overall time index (OTI), expressed as a percentage, can be defined as a difference of a ratio of non-picking to picking time percentage from one (Eq. 8.2). Similarly, from Eq. (8.2) it follows that a reduced value of non-picking and increased value of picking time will improve the OTI. The OE and OTI, with their component parameters, could reflect the harvest efficiency in an objective manner.

$$\text{Overall efficiency (OE, \%)} = \frac{\text{Picking time percentage (PTP, \%)} \times \text{Detaching time percentage (DTP, \%)}}{100} \tag{8.1}$$

$$\text{Overall time index (OTI, \%)} = \left(1 - \frac{\text{Non} - \text{picking time percentage}}{\text{Picking time}}\right) \times 100 \quad (8.2)$$

The OEs for the ladder-and-bucket, DBR, and Huron methods are 21%, 23%, and 26%, respectively (Fig. 8.7). The Huron performs better than the other two methods and is mainly because workers are free from handling ladders, moving platforms, as well as workers are not involved with machine driving. Compared to ladder-and-bucket method, DBR and Huron improved the OEs by 10% and 24%, respectively. Considering the relatively low values of OEs among all harvesting methods considered, in general, there is good scope for improvement (>76%), which can be addressed in the future through research and development of innovative approaches to apple harvesting. Whereas the results of OTI of 45%, 71%, and 80% for the ladder-and-bucket, DBR, and Huron methods, respectively, indicate the conventional method of ladder-and-bucket has huge potential (55%) for improvement while the modern DBR (29%) and Huron (20%) methods demonstrating good OTIs have limited but definite potential for improvement. Therefore, the results of this study identified a great opportunity for improvement, and it is certain that future efforts on apple harvesting could achieve significant progress in terms of OE and OTI.

One potential method is to develop a catching mechanism and mount it right below the hand level of the workers, wherein workers only need to detach apples and the catching mechanism would complete the transportation work. With such a catching mechanism, the apple approaching time percentage is also saved partially as the hands will stay more in the picking zoon. The study results indicate that more efforts should be directed to the improvement of apple picking component activities to improve the overall apple harvesting process as well as improving the harvesting methods through innovative design and development.

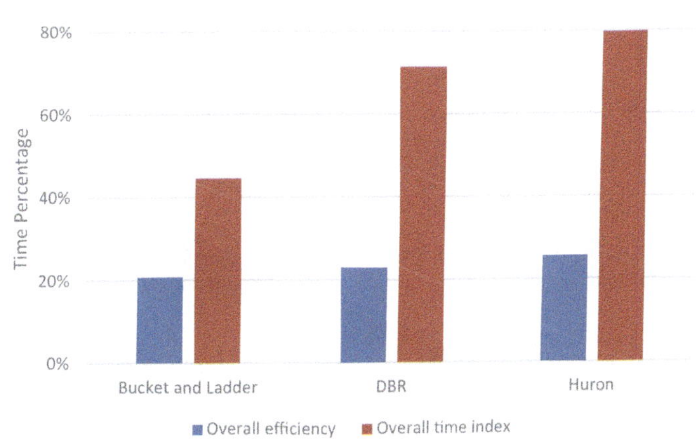

Fig. 8.7 Overall efficiency and overall time index among different apple harvest methods showing the potential scope for improvement

8.4 Conclusion

This study was able to categorize the continuous apple harvesting process from the video recording analysis into different activities for the three different methods, such as ladder-and-bucket, DBR, and Huron. Harvesting activities were re-grouped into picking and non-picking and the time percentage for each activity was calculated by analyzing the extracted video images. Picking time percentage (PTP) results indicated that pickers working with ladder-and-bucket, DBR, and Huron spent 64%, 78%, and 83%, respectively, of their total time in picking apples. Workers using the Huron platform had the highest PTP because they are free from ladder use, moving the platform, and driving the machine. Further evaluation of the picking process in terms of approaching, detaching, and transporting apples showed that workers only spent ~ 1/3 of the picking time in detaching apples for all the three methods, which meant a majority of picking time (~2/3) was spent on approaching and transporting actions of the hands. Overall efficiency of 21%, 23%, and 26%, and overall time index of 45%, 71%, and 80% for ladder-and-bucket, DBR, and Huron methods, respectively indicated that the DBR and Huron achieved an increase of 10% and 24% of overall efficiency and 58% and 78% of overall time index over the ladder-and-bucket method.

Considering the PTP has already been in a reasonable high range (64% to 83%) as well as the higher values of overall time index (71% to 80%) for the modern DBR and Huron methods, there is less scope for their improvement but a great opportunity for ladder-and-bucket method. However, the low values of detachment time percentage (1/3 of entire picking time) and the existing gap based on the overall efficiency and time index results indicated a clear potential for improvement for all the harvest methods studied. One potential method is developing a catching mechanism that would complete the transportation of apples thereby freeing the workers from transporting and partially saving on the hand approaching activity. It is therefore concluded that more research efforts should be directed to the improvement of apple picking component activities, the overall harvesting process, and the harvesting methods through innovative design and development.

Credit Authorship Contribution Statement **Z. Zhang**: Writing—original draft. **Z. Zhang**: Conceptualization, Investigation, Methodology, Project administration, Writing—original draft, Writing—review & editing.

Disclaimer
Mention of commercial products or orchards in this paper is only for providing factual information and does not imply endorsement of them by authors over those not mentioned.

Declaration of Competing Interest
The authors declare that they have no known competing financial interests or personal relationships that could have appeared to influence the work reported in this paper.

References

1. Boyer J, Liu RH (2004) Apple phytochemicals and their health benefits. Nutr J 3(1):5
2. He L, Fu H, Karkee M, Zhang Q (2017) Effect of fruit location on apple detachment with mechanical shaking. Biosys Eng 157:63–71
3. Zhang Z, Heinemann P, Liu J, Baugher T, Schupp J (2016a) Development of mechanical apple harvesting technology—a review. Trans ASABE 59(5):1165–1180
4. Pothula AK, Zhang Z, Lu R (2018) Design features and bruise evaluation of an apple harvest and in-field presorting machine. Trans ASABE 61(3):1135–1144
5. Zhang Z, Heinemann P, Liu J, Baugher T, Schupp J (2016b) Design and field test of a low-cost apple harvest-assist unit. Trans ASABE 59(5):1149–1156
6. Freivalds A, Park S, Lee C, Earle-Richardson G, Mason C, May JJ (2006) Effect of belt/bucket interface in apple harvesting. Int J Ind Ergon 36(11):1005–1010
7. Lesser K, Harsh RM, Seavert C, Lewis K, Baugher T, Schupp J, Auvil T (2008) Mobile platforms increase orchard management efficiency and profitability. In: International symposium on application of precision agriculture for fruits and vegetables, pp 361–364
8. Zhang Z, Pothula A, Lu R (2017a) Economic evaluation of apple harvest and infield sorting technology. Trans ASABE 60(5):1537–1550
9. Fathallah FA (2010) Musculoskeletal disorders in labor-intensive agriculture. Appl Ergon 41(6):738–743
10. Zhang Z, Heinemann P, Liu J, Schupp J, Baugher T (2017c) Brush mechanism for distributing apples in a low-cost apple harvest-assist unit. Appl Eng Agric 33(2):195–201
11. Zhang Z, Pothula A, Lu R (2017b) Development and preliminary evaluation of a new bin filler for apple harvesting and infield sorting machine. Trans ASABE 60(6):1839–1849
12. Earle-Richardson G, Jenkins PL, Strogatz D, Bell EM, May JJ (2006) Development and initial assessment of objective fatigue measures for apple harvest work. Appl Ergon 37(6):719–727
13. Earle-Richardson G, Jenkins P, Fulmer S, Mason C, Burdick P, May J (2005) An ergonomic intervention to reduce back strain among apple harvest workers in New York State. Appl Ergon 36(3):327–334
14. Sakakibara H, Miyao M, Kondo TA, Yamada SY (1995) Overhead work and shoulder-neck pain in orchard farmers harvesting pears and apples. Ergonomics 38(4):700–706
15. Zhang Z (2015) Design, test, and improvement of a low-cost apple harvest-assist unit. Ph.D. diss. University Park: Pennsylvania State University, Dept Agric Biol Eng
16. Zhang Z, Heinemann P (2017) Economic analysis of a low-cost apple harvest-assist unit. HortTechnology 27(2):240–247
17. Lu R, Pothula AK, Mizushima A, VanDyke M, Zhang Z (2018) U.S. Patent No. 9,919,345. U.S. Patent and Trademark Office, Washington, DC
18. Lu R, Zhang Z, Pothula A (2017) Innovative technology for enhancing apple harvest and postharvest handling efficiency. Fruit Q 25(2):11–14
19. Schupp J, Baugher T, Winzeler E, Schupp M, Messner W (2011) Preliminary results with a vacuum assisted harvest system for apples. Fruit Notes 76(4):1–5
20. Davidson JR, Mo C, Zhang Q, Silwal A, Karkee M (2017) U.S. Patent No. 9,554,512. Washington, DC: U.S. Patent and Trademark Office
21. Davidson JR, Silwal A, Hohimer CJ, Karkee M, Mo C, Zhang Q (2016) Proof-of-concept of a robotic apple harvester. In Intelligent Robots and Systems (IROS), IEEE/RSJ Int Conf 634–639
22. Baeten J, Donné K, Boedrij S, Beckers W, Claesen E (2008) Autonomous fruit picking machine: a robotic apple harvester. Field and service robotics. Springer, Berlin, pp 531–539
23. Stuntz S (2018) Canopy preparation helps robotic apple harvester. Fruit Growers News. Accessed from https://fruitgrowersnews.com/article/canopy-preparation-helps-robotic-apple-harvester/
24. Zhao D, Lv J, Ji W, Ying Z, Yu C (2011) Design and control of an apple harvesting robot. Biosys Eng 110(2):112–122
25. Silwal A, Davidson JR, Karkee M, Mo C, Zhang Q, Lewis K (2017) Design, integration, and field evaluation of a robotic apple harvester. J Field Robot 34(6):1140–1159

26. Kahani A, Dror B (2016) U.S. Patent No. US 9,475,189 B2. U.S. Patent and Trademark Office, Washington, DC
27. Dininny S (2017) The latest on FF Robotics' machine harvester. Good Fruit Growers. Accessed from https://www.goodfruit.com/the-latest-on-ff-robotics-machine-harvester/
28. Peterson DL (2005) Development of a harvest aid for narrow- inclined-trellised tree-fruit canopies. Appl Eng Agric 21(5):803–806
29. Peterson DL, Miller SS (1996) Apple harvesting concepts for inclined trellised canopies. Appl Eng Agric 12(3):267–271
30. Zhang Z, Pothula A, Lu R (2018) A review of bin filling technologies for apple harvest and postharvest handling. Appl Eng Agric 34(4):687–703
31. Zhang Z, Pothula A, Lu R (2019) Improvements and evaluation of an infield bin filler for apple bruising and distributions. Trans ASABE 62(2):271–280
32. Baumgart A, Neuhauser D (2009) Frank and Lillian Gilbreth: scientific management in the operating room
33. Gilbreth FB (1914) Scientific management in the hospital. Mod Hosp 3:321–324
34. Taylor FW (1914) The principles of scientific management. Harper
35. Hendrich A, Chow MP, Skierczynski BA, Lu Z (2008) A 36-hospital time and motion study: how do medical-surgical nurses spend their time? Permanente J 12(3):25
36. Lan S, Wang X, Ma L (2009) Optimization of assembly line based on work study. In: 16th international conference and industrial engineering and engineering management, IE&EM'09, pp 813–816
37. Duran C, Cetindere A, Aksu YE (2015) Productivity improvement by work and time study technique for earth energy-glass manufacturing company. Procedia Econ Finance 26:109–113
38. Moktadir MA, Ahmed S, Zohra FT, Sultana R (2017) Productivity improvement by work study technique: a case on leather products industry of Bangladesh. Ind Eng Manag 6:1–11

The manufacturer's authorised representative in the EU is Springer Nature Customer Service Centre GmbH, Europaplatz 3, 69115 Heidelberg, Germany. If you have any concerns regarding our products, please contact ProductSafety@springernature.com

Printed and bound by CPI Group (UK) Ltd, Croydon, CR0 4YY
30/09/2026
02225259-0008